Single-level & Underground Home Plans

Single-level & Underground Home Plans

Edited by Garlinghouse

THE GARLINGHOUSE COMPANY
Topeka, Kansas

SINGLE-LEVEL & UNDERGROUND HOME PLANS

A Garlinghouse Company publication.

ISBN 0-938708-01-5

Printed in the United States of America.

Library of Congress catalog card number 81-81751.

IF RISING FUEL COSTS MAKE YOU BLUE, THINK PINK.

Cheer up, things could get worse.

The cost of fuel, for instance. It may double in the next ten years.*

But you can fight back. With an extra layer of pink Owens–Corning *Fiberglas®* insulation in your attic. Pink Owens–Corning is America's best-selling insulation. Add a little extra.

Insulate your attic, walls exposed to heat and cold, and floors over unheated garages and crawl spaces. The way things are, it could be more expensive *not* to.

Savings vary. Find out why in the sellers fact sheet on R-values. Higher R-values mean greater insulating power.

OWENS/CORNING
FIBERGLAS
TRADEMARK ®

Going Underground!

Building underground to save on long-term maintenance and fuel costs does not have to mean that you will have to give up all thoughts of a bright, sunny structure with a breath-taking view. In fact, underground homes being built across the country have some of the most impressive views in their area. This is because, even though these buildings have earth-covered roofs and protecting earth at most outside walls, they are not necessarily deep below grade. Today, there are a number of possibilities for underground construction and several considerations to make to avoid costly mistakes.

Taking time to select just the right design for your site is critical. After all, you're creating a structure that will affect thousands of dollars and change the face of the land for many lifetimes. It's best to stick to a simple design and eliminate wasted space, complicated roofs and expensive gadgetry. You'll get better results with a simple, compact and well-planned building.

The water table in your area is of primary consideration when considering an underground structure for it is essential to build above it. You should determine the highest level that ground water may reach at you building site by enlisting the aid of a soils engineer or the Soil Conservation Service.

Building codes should be another major consideration. The Uniform Building Code, for instance, requires operable windows or outside doors in bedrooms for fire exits. Before building, it's wise to check out local building codes for other requirements.

Since a major reason for considering an underground structure is fuel conservation, it makes sense to look out for possible heat leaks. You can conserve on possible heat leaks by steering away from the cantilever roof slabs and parapet walls which were popular in the first generation of underground buildings. Skylights should also be carefully considered. Not only are they potentials for water leaks, but they have a way of letting heat pour out in the winter and in the summer. A better approach is roof monitors with south-facing glazing to replace the plastic dome.

Insulation is critical in avoiding major heat loss and earth will not do the job by itself. Earth is great at moderating temperature around a building and protecting it from viscious winds but insulation should be used as well.

Materials used in underground construction must be substantial to avoid future digging to make repairs. Superior products, even though they are costly, will last longer. You should use only waterproofing products which the manufacturer specifically recommends, in literature, for underground use. Most any waterproofing application must be treated as a "system" using the primers, edge treatments and accessories as recommended by the manufacturers. Whatever material you use, the crucial factor is the quality of workmanship during installation. A careful-do-it-yourself job might even be better than a contractor's work.

Rigid insulation boards all around the structure and in contact with earth must be selected carefully so that they will not eventually soak up water. Check manufacturer's literature to be sure a product is recommended for soil contact and use in water-logged conditions.

Humidity levels don't have to be a problem in properly designed underground structure if you remember a few basics. In a new building, curing concrete releases moisture into the interior air which makes a dehumidifier necessary for perhaps the first two years as the structure dries out.

Landscaping around the structure is another important consideration. A reasonable depth of earth cover on the roof often ranges from one and a half feet to two feet. It is important to grade all slopes away from the building to avoid ponding and undue water pressure. Deep mulching on top of the soil provides a lightweight covering that helps retain soil moisture, prevents erosion and provides a protective cover for young plants. Rooftop plantings may include some transplanted native shrubs and ground covers with the major portion of the roof allowed to go through natural succession that determines what's best suited for that particular place.

The real delight about building an underground structure is its way of becoming part of the environment. You will find that your underground structure will change for the better as years go by. As wildlife makes itself at home on top of and around your building, its early unfinished appearance will slowly be changed by nature.

Contents

Energy Efficient Plan Uses Berms

No. 10358—Semi-underground to save energy, this handsome design calls for soil bermed up to the cornices on sides and back and an airlock type vestibule. Plenty of warm sunlight penetrates the well-windowed front, designed to face south, and rear and side window wells allow for emergency exits. For family-oriented livability, the great room with wood-burning fireplace merges with the dining area, and a functional corridor kitchen is handy to both. Located off the double garage, the mud room functions as a laundry and utility area. Three large bedrooms share two full baths.

First floor—1,620 sq. ft.
Garage—448 sq. ft.

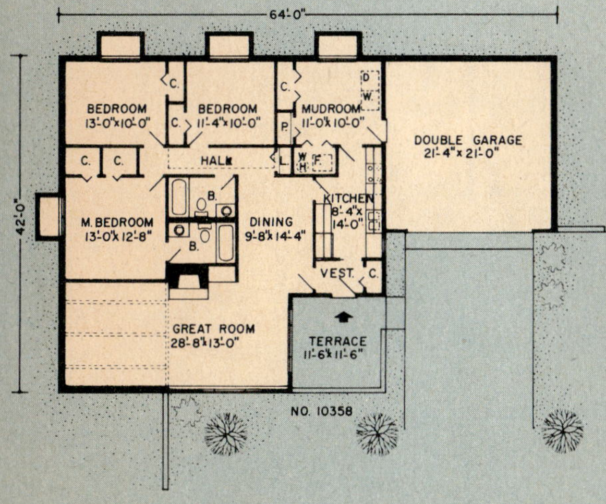

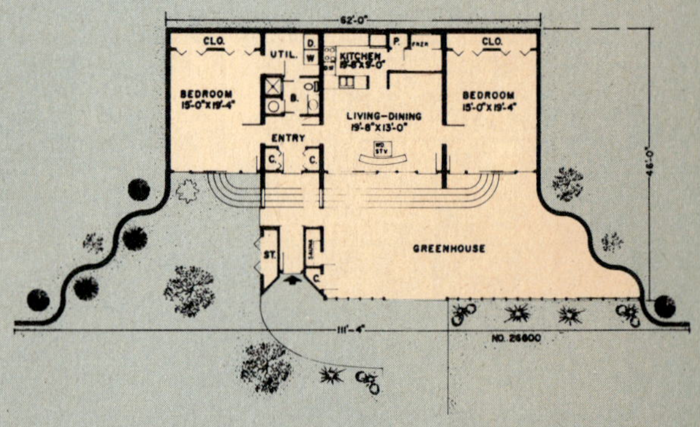

Underground With Greenhouse

No. 26600—This quality home shows many delightful features designed to conserve energy and provide lots of enjoyment. The thoughtfully planned kitchen and the spacious living/dining room combination are centrally located with sliding glass doors opening to the large greenhouse complete with sauna. A large bedroom on each end provides plenty of privacy and both have sliding glass doors, one exiting to the greenhouse and the other to the patio. Storage space abounds in this lovely underground home.

First floor—1,609 sq. ft.
Greenhouse—910 sq. ft.

*For price and order information
see pages 108-109.*

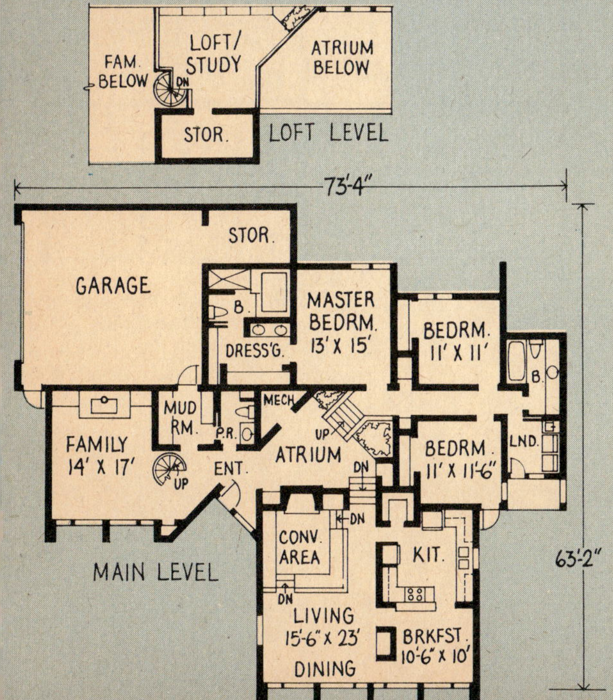

Three Bedrooms And A Loft

No. 26890—Words alone cannot adequately describe this marvelous three bedroom home. The main floor is on three levels with kitchen, breakfast nook, dining area and conversation pit on the lower level. Upon entering this home, you are treated to a view of the well designed atrium and then to your left is a family room. The third level shows two large bedrooms sharing a bath and a centrally located laundry room. The master bedroom is also on this level and is lavishly complimented with a private bath which includes it's own whirlpool and large dressing room. Completing the picture is a spiral stair leading to the loft which overlooks the atrium. Earth berming contributes to the energy efficient qualities of this fine home.

2,744 sq. ft.

DESIGNER: Charles Woods

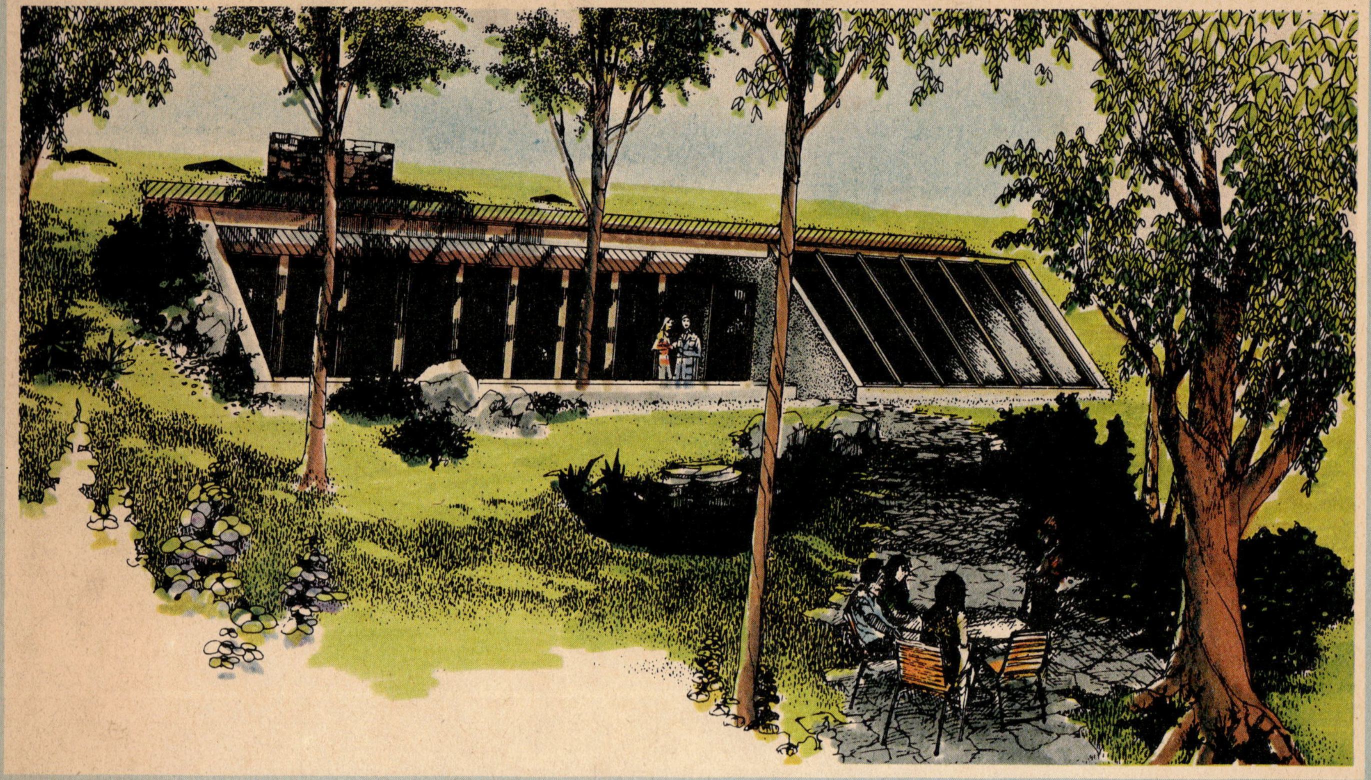

Energy Savings And Beauty

No. 26050—An unbeatable combination. This passive solar underground two bedroom home offers the unique ability of combining two energy saving features with beauty, grace and style. Highlighting this home is a sunken court yard and an attached greenhouse/sunroom, for that summer feeling in the winter. The foliage also keeps the climate in this well sealed home fresh. A skylight floods the spacious kitchen with sunlight and a central fireplace seperates the living room from the family/dining area, allowing for an open spacious feeling. A study and well designed utility room completes this excellently zoned floor plan.

1,934 sq. ft.

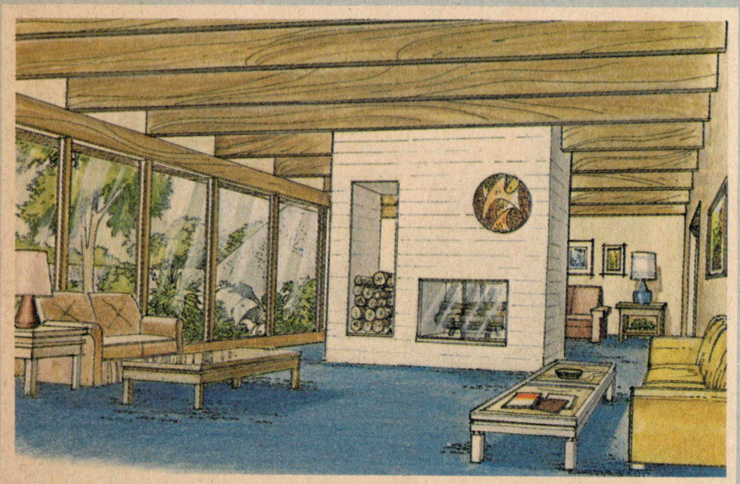

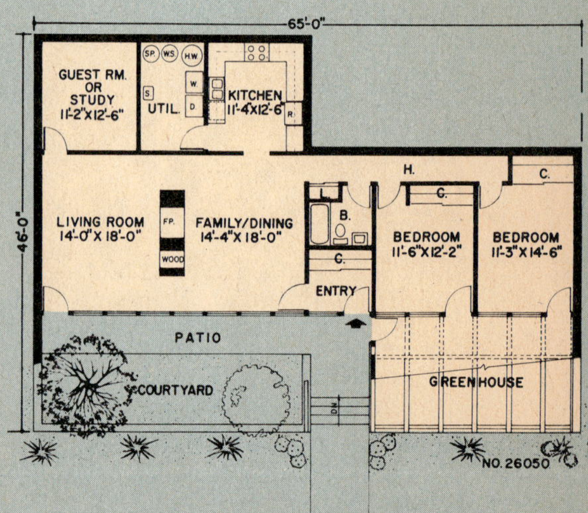

For price and order information see pages 108-109.

Underground Home Cuts Energy Use

No. 10364—Oriented to a southern exposure, this underground home combines all the comforts of a well-planned contemporary with a fraction of the typical energy consumption. The pre-stressed concrete roof is covered with two feet of earth, a highly effective insulator, to keep winter cold out, warmth in. Walls on three sides are of 8-inch poured concrete. In addition to this functional approach, the plan is highly livable

. . . with a protected patio reached by sliding glass doors and an immense open living area featuring family room, living room, and kitchen with snack bar. Bedrooms are large, and the master bedroom annexes a private bath and three closets. A utility/laundry room, double garage, and chimneys for gas furnace and wood stove are included.

House—1,795 sq. ft.
Garage—576 sq. ft.

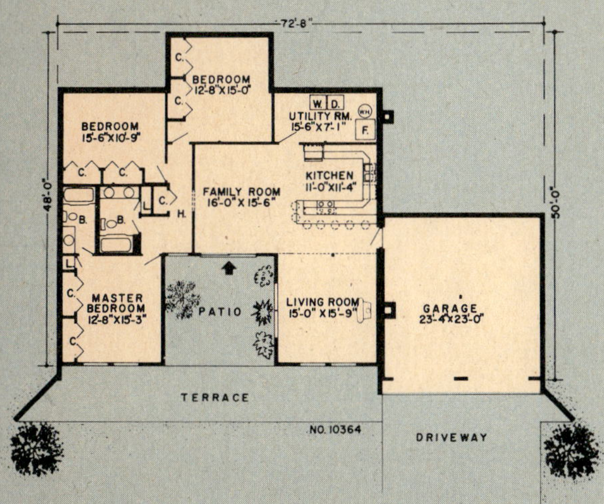

Underground Delight

No. 10376—This three bedroom, underground masterpiece is designed to fight the high cost of living through its many energy saving features, including the use of passive solar energy. The large master bedroom on one end shows an abundance of closets. Each of the three bedrooms has sliding glass doors to the front lawn. Also featured in this area is a multi-purpose room, easily converted to individual use, graciously seperated from the entry way by ornately carved wood room dividers. The plan calls for 2 baths, one delightfully designed with a whirlpool. The family room opens to a greenhouse via sliding glass doors. A two car garage completes this home.

2,086 sq. ft.

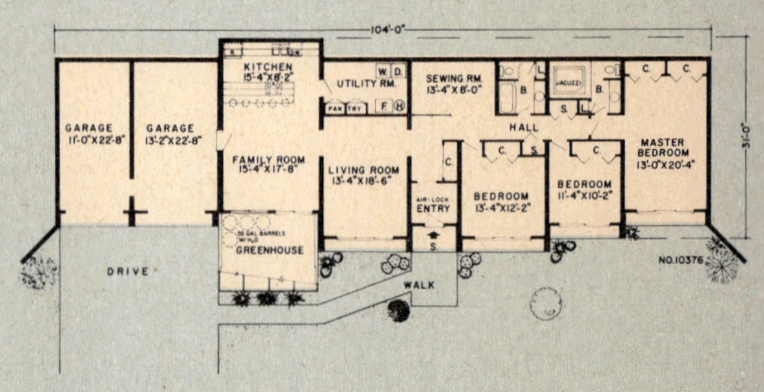

*For price and order information
see pages 108-109.*

Home Plans You Can Buy

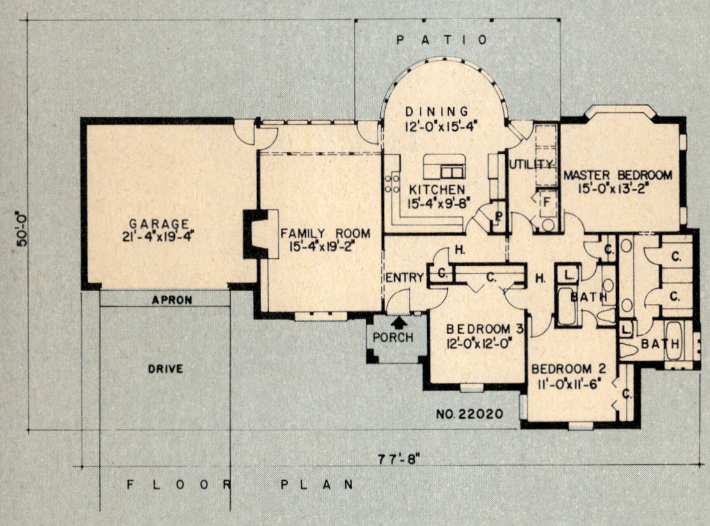

Tasteful elegance aim of design

No. 22020—With an exterior that expresses French Provincial charm, this single level design emphasizes elegance and offers a semi-circular dining area overlooking the patio. To pamper parents, the master bedroom annexes a long dressing area and private bath, while another bath serves the second and third bedrooms. A wood-burning fireplace furnishes the family room.

House proper—1,772 sq. ft., Garage—469 sq. ft.

For price and order information
see pages 108-109.

Charming traditional emphasizes living areas

No. 22014—Besides its 20-ft. family room with fireplace, this one story traditional calls for a dining room, breakfast nook, and sizable gameroom that can function as a formal living room if preferred. Each of the three bedrooms adjoins a full bath, with the master bedroom meriting a luxurious "his and hers" bath with two walk-in closets.

House—2,157 sq. ft., Garage—485 sq. ft.

*For price and order information
see pages 108-109.*

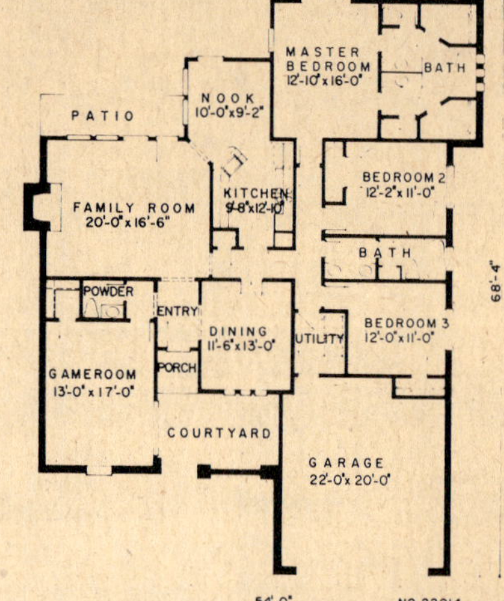

Courtyard adds interest to plan

No. 22010—Well-defined contemporary lines are softened by a semi-enclosed courtyard visible from the dining area of this striking design. The 30-ft. family room is dominated by a fireplace, resulting in a spacious but cozy area for entertaining. The island kitchen merges with dining nook, and bedrooms are large, featuring the master bedroom and its luxurious bath.

House—2,174 sq. ft., Garage—506 sq. ft.

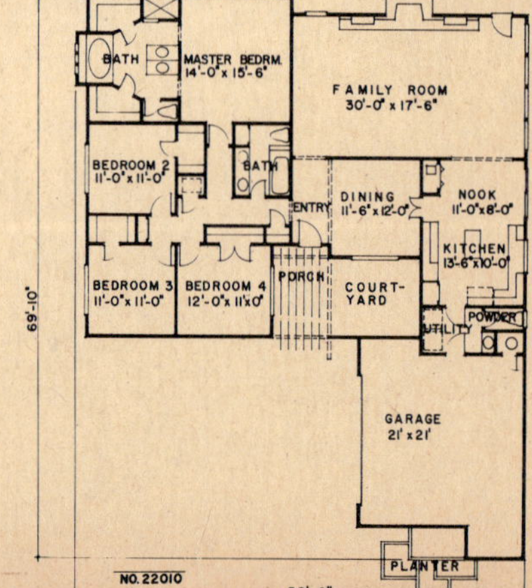

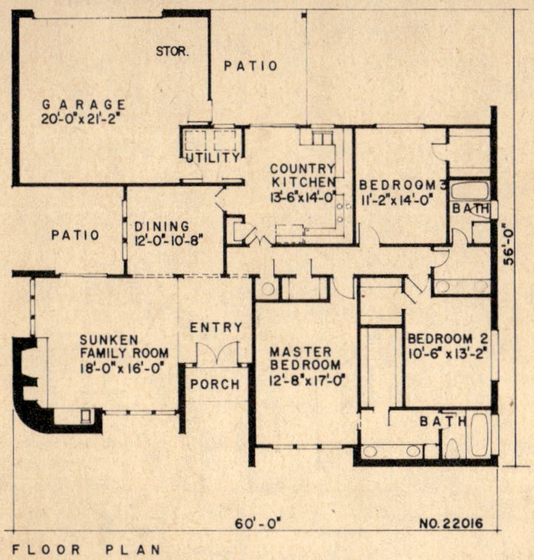

Outstanding plan shows sunken family room

No. 22016—Contemporary in its approach to family living, this single level design equips its sunken family room with wood-burning fireplace and sliding glass doors to the patio. Country kitchen and garage are separated by a laundry/mud room, and a formal dining room is offered. Notable is the master bedroom, with its large areas of wall space and private bath with walk-in closet.

House—1,880 sq. ft., Garage—524 sq. ft.

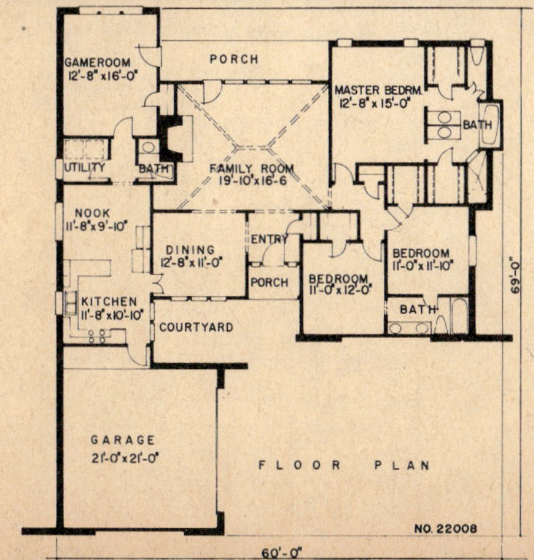

Hip roof design, family-centered space

No. 22008—Inside this trim hip roof plan, space is allotted for a variety of family activities. Spotlighted is the sizable beamed family room with fireplace and access to porch. The bordering gameroom edges a handy half bath, and the dining nook connects to, and visually enlarges, the kitchen. Four bedrooms and two full baths are planned.

House—2,074 sq. ft., Garage—544 sq. ft.

Brick-layered home plans 4 bedrooms

No. 22004—Four roomy bedrooms, featuring a master bedroom with extra large bath, equip this plan for a large family or overnight guests. The centrally located family room merits a fireplace, wet bar, and access to the patio, and a dining room is provided for formal entertaining. An interesting kitchen and nook, as well as two and one half baths, are featured.

House—2,070 sq. ft., Garage—474 sq. ft.

*For price and order information
see pages 108-109.*

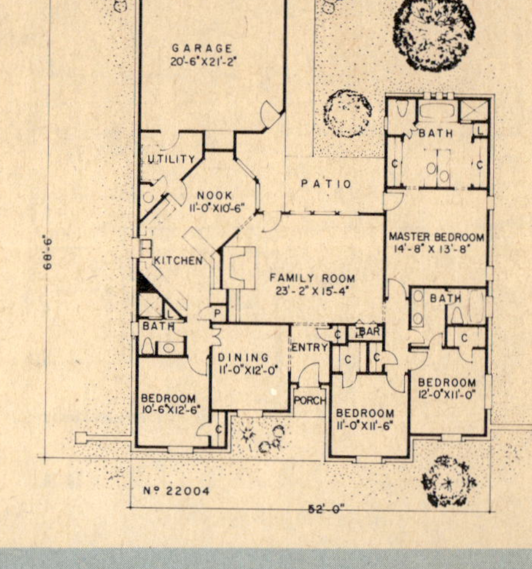

Exterior charm, interior space

No. 21500—Brick trim and diamond lite windows enhance the exterior of this engaging one-story traditional. Inside, the emphasis is on space. Besides four large bedrooms, the plans call for a 22 ft. family room that opens to a covered porch, a game room with built-in bar sink and large kitchen with pantry and breakfast nook. Vaulted ceilings are featured in the master bedroom, which adjoins a private bath and dressing room.

House—2,627 sq. ft., Garage—527 sq. ft.

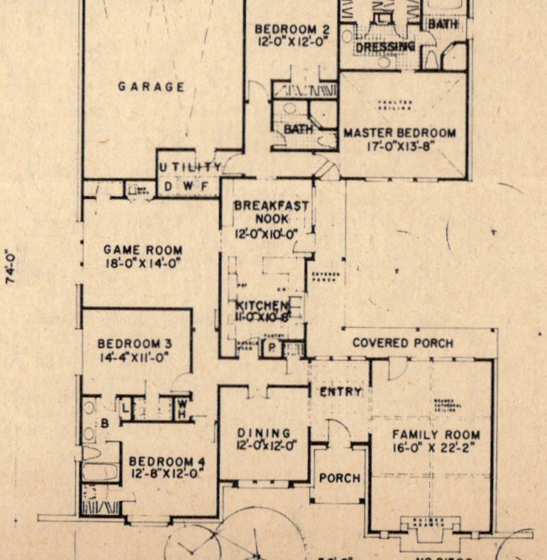

Unique design boasts solarium

No. 21502—Tucked between the entry and the family room of this exceptional four bedroom plan, the solarium is the focal point of the home. Three baths are well placed, off the master bedroom, between the second and third bedrooms and near the fourth bedroom and family room. A spacious formal dining room and side entry garage are provided.

House—2,270 sq. ft., Garage—473 sq. ft.

Courtyard sets pace for home

No. 21506—Refreshing in concept, this one story design is reached via a long courtyard that sets a gracious tone. Master bedroom which merits a private bath and dressing area is flanked by two patios. For convenience, a utility room separates the double garage and kitchen. A sizable family room is included.

House—1,737 sq. ft., Garage—452 sq. ft.

Terrace focus of contemporary plan

No. 380—Sizable kitchen and family room adjoin the large terrace, creating a center for picnics and entertaining in this efficient one story plan. Living and dining rooms offer a formal contrast and are readily accessible from the front entry. Four bedrooms occupy the right wing of the design and include a front-facing master bedroom meriting double closets and full bath. Large basement promises space for recreation, and laundry space is offered in the kitchen.

First floor—1,873 sq. ft., Basement—1,660 sq. ft.

*For price and order information
see pages 108-109.*

Zoning effective in contemporary

No. 10152—Formal living room and foyer serve to zone quiet and active areas in this thoughtfully planned contemporary. Comprising three bedrooms and three full baths, the home displays a floor plan geared to family activities and adaptable to entertaining. Living and family rooms each enjoy wood-burning fireplaces and sliding glass doors to the expansive patio. At front, the kitchen is square and spacious and is flanked by formal dining room and laundry, with storage closet and full bath steps away.

**First floor—1,816 sq. ft., Basement—1,816 sq. ft.
Garage—576 sq. ft.**

Fireplace adds warmth

No. 358—The utility of a well-proportioned living room which opens onto a spacious concrete terrace highlights this attractive house. Over 135 square feet of kitchen space, conveniently situated between laundry and dining rooms will delight efficient housewives as will the plentiful closet and storage space. Three bedrooms, two full baths and a double garage complete this ranch style home, designated to win the admiration of friends and neighbors.

First floor—1,683 sq. ft., Basement—1,683 sq. ft.
Garage—453 sq. ft.

Firelight enriches living, entertaining

No. 10136—Relaxed suburban living is the aim of this rambling traditional, highlighted by a striking three-way fireplace that lights and separates living and dining rooms. Expanses of sliding glass doors allow the area to share the terrace, and the corridor kitchen at right offers help in entertaining. Open to the kitchen is a useful complex which includes laundry, family room, and a closeted full bath near garage and basement work areas for ease of clean-up.

First floor—1,888 sq. ft., Basement—1,184 sq. ft.
Garage—590 sq. ft.

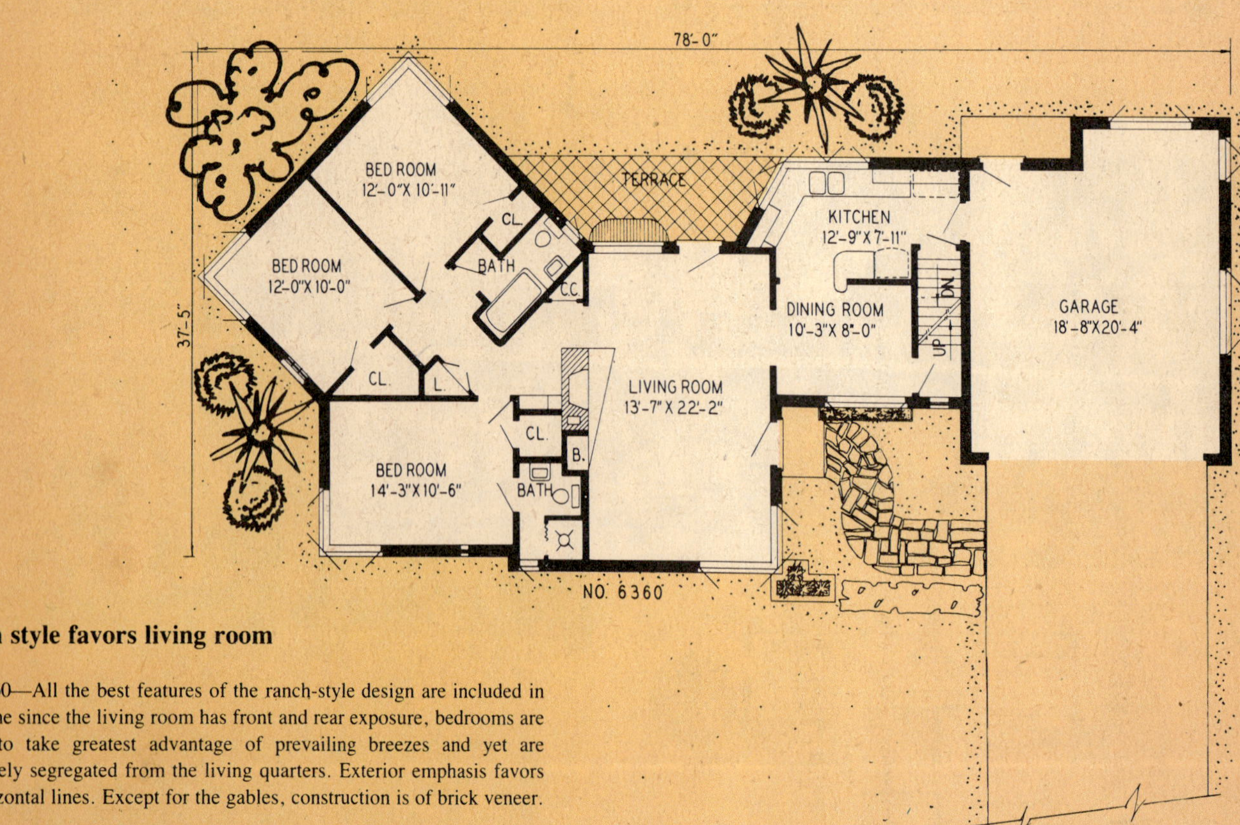

Ranch style favors living room

No. 6360—All the best features of the ranch-style design are included in this home since the living room has front and rear exposure, bedrooms are placed to take greatest advantage of prevailing breezes and yet are completely segregated from the living quarters. Exterior emphasis favors the horizontal lines. Except for the gables, construction is of brick veneer.

**House—1,293 sq. ft., Basement—767 sq. ft.
Garage—466 sq. ft., Terrace—92 sq. ft.**

*For price and order information
see pages 108-109.*

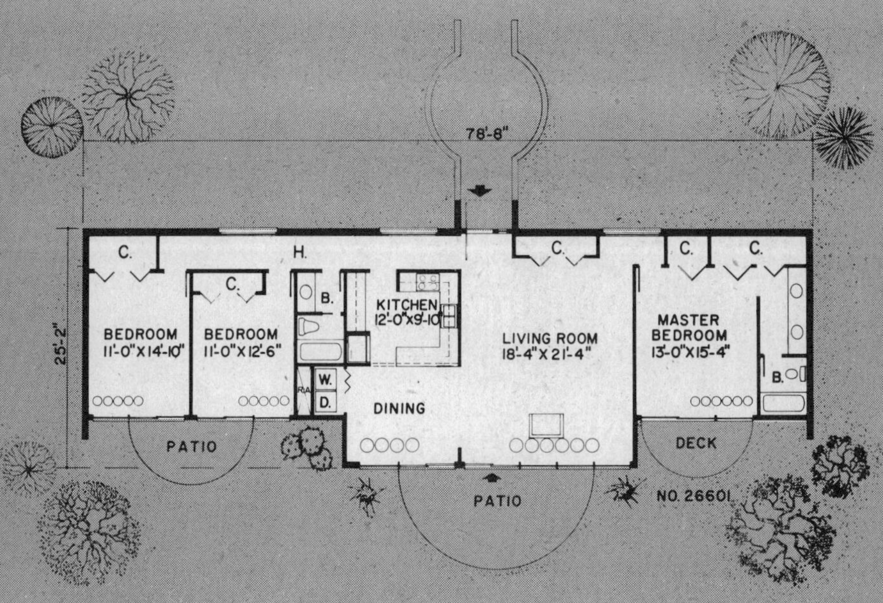

Fight the high cost of energy

No. 26601—This striking contemporary passive solar home, designed for a growing family, can help combat run-away fuel bills. Large expanses of glass on the south side warms the house during the day while storing heat in solar water tubes positioned behind windows for heat at night. Master bedroom has private bath and a redwood deck. Living room, dining room and kitchen are the central attractions in this well zoned home. Sliding glass doors give access to the patio from both the living room and dining room. Two bedrooms share a patio accessible from each room via sliding glass doors. The north side of the house is bermed up with earth for protection from cold north winds. Super insulation in walls and ceiling complete this energy efficient, economical home.

Living area—1,748 sq. ft.

Trim plan designed for handicapped

No. 10360—Attractive and accessible, this three bedroom home has been carefully detailed to provide both
comfort and self-sufficiency for the handicapped individual. Ramps allow entry to garage, patio and porch.
Doors and windows are located so that they can be
opened with ease, and both baths feature wall-hung
toilets at a special 16-18" height. Spacious rooms,
wide halls, and the over-sized double garage allow a
wheelchair to be maneuvered with minimal effort, and
the sink and cooktop are also located with this in
mind. Besides these functional aspects, the design
also boasts a great room, inviting and open to the
living areas and patio and supplied with a wood-
burning fireplace. The sizable master bedroom merits
an outside entry, both a convenience and a safety
measure, and adjoins a private bath with large shower
accessible by wheelchair.

First floor—1,882 sq. ft., Garage—728 sq. ft.

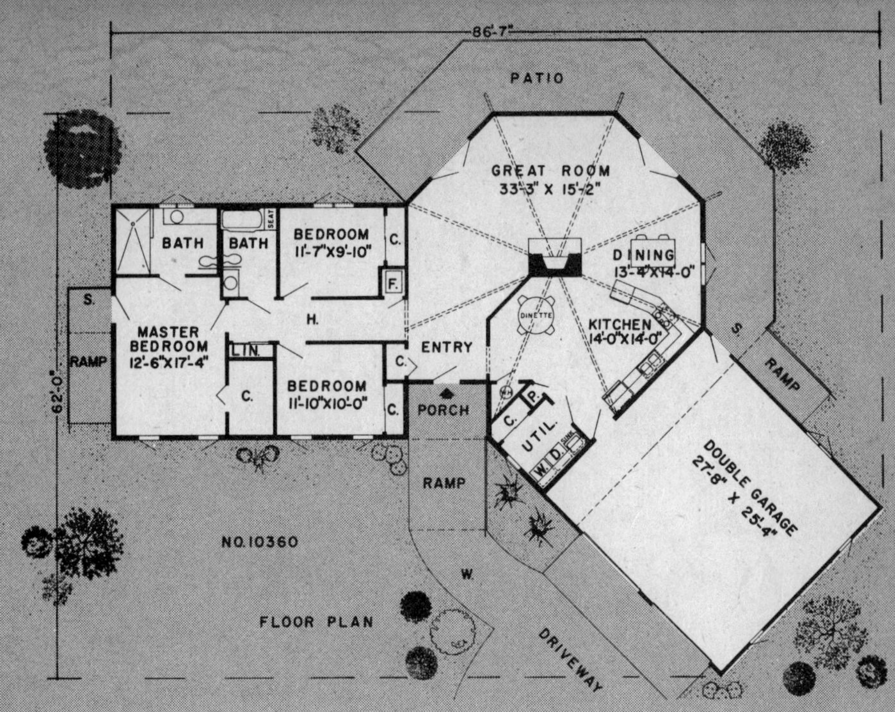

Roofed terrace separates garage, home

No. 19691—For charm and convenience, a roofed
terrace links the garage with this well-arranged ranch
style plan. A large living room with fireplace merges
with the dining area, and an efficient kitchen includes
a pantry. Bordering the kitchen, a nook allots space
for dining as well as laundry and sewing equipment.
Three bedrooms and two baths, one with sunken tub,
are featured.

First floor—2,136 sq. ft., Garage—644 sq. ft.

*For price and order information
see pages 108-109.*

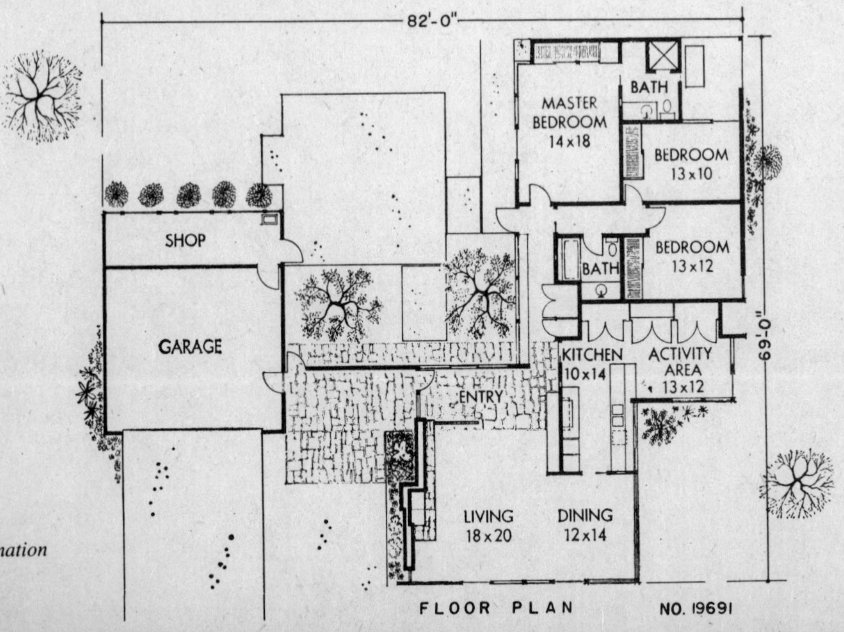

Design calls for formal living areas

No. 19718—Shuttered small-paned windows accent a traditional exterior, while on the inside, formal living and dining rooms balance the paneled informality of the family room. For convenience, the kitchen borders a breakfast room and extra large utility room with space for laundry equipment and freezer. Featured is the private study with built-in bookshelves.

Living area—2,421 sq. ft., Garage—560 sq. ft.

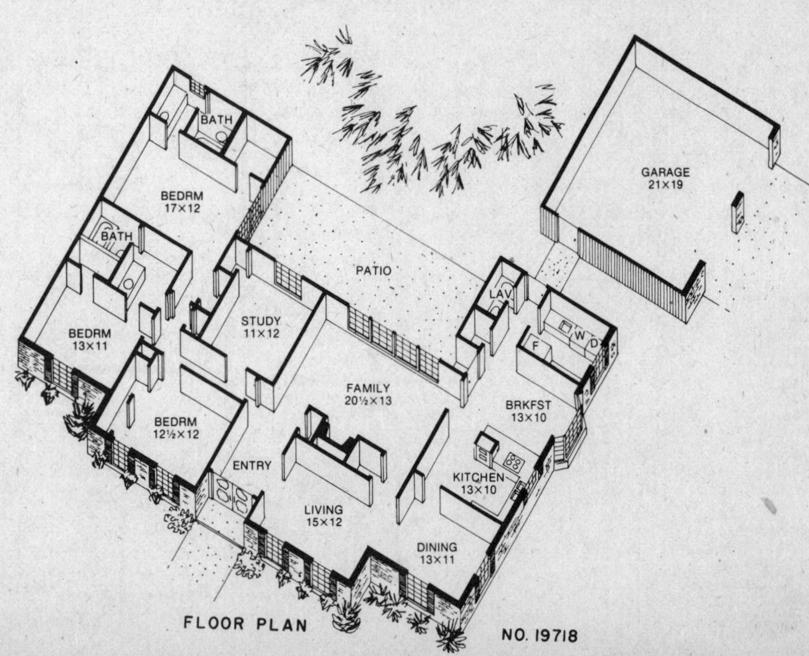

Plan suits family with children

No. 1004—Spacious family room, out of the way and opening to patio, supplies a quiet haven for children's play or television viewing in this brick-layered design. For adult entertaining, the elongated living-dining room is within a few feet of entry and kitchen. Parents' bedroom is indulged with a corner of its own and a private bath with shower and linen closet. Storage space is furnished in the extra large garage and laundry space in the eat-in kitchen.

First floor—1,588 sq. ft., Garage—480 sq. ft.

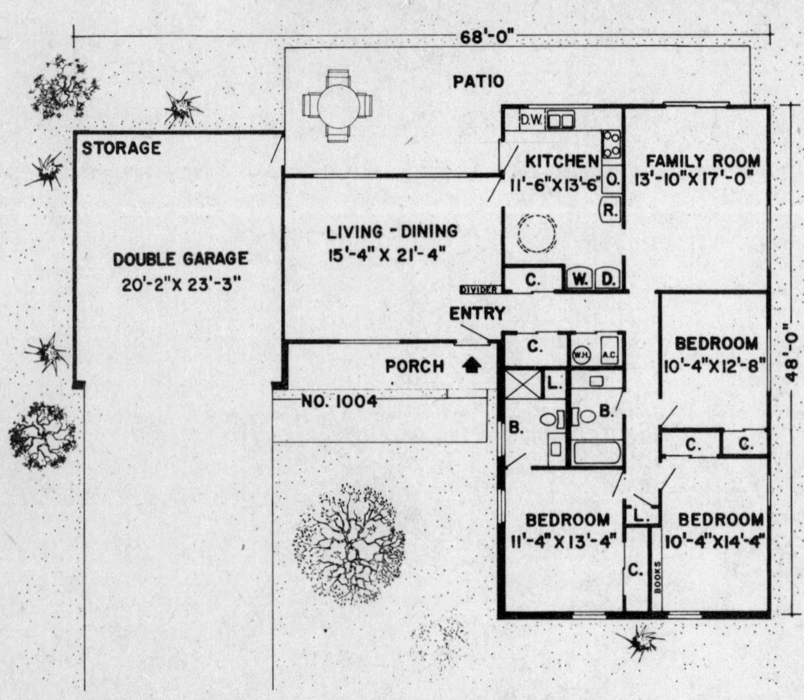

For price and order information see pages 108-109.

Courtyard plan given contemporary slant

No. 10130—Sprawling and lavish, this unusual contemporary plan displays a walled courtyard and surrounds a patio and atrium with pool. Sliding glass doors open the atrium to all sections of the home and create a pleasant fusion of indoors and outdoors. Views of the pool are especially enjoyed by beamed, firelit living room at right and large, eat-in kitchen above. Three bedrooms border the long, compartmented bath, and a closed-off laundry center adjoins the kitchen.

First floor—1,734 sq. ft.
Patio & atrium—292 sq. ft., Garage—592 sq. ft.

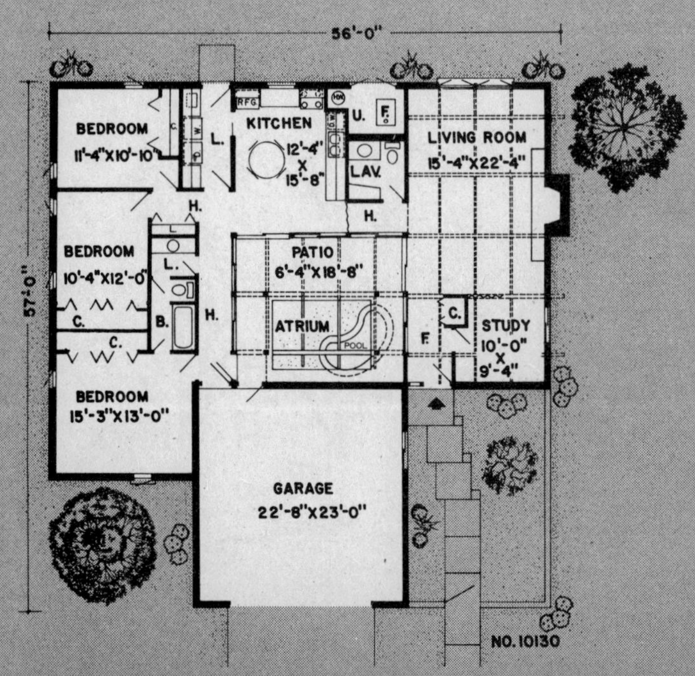

Open plan asset in distinctive home

No. 10144—Expansive and spacious, this stone-trimmed four bedroom design uses open planning to its fullest advantage. Rooms, even bathrooms, are large, and dining and family rooms increase space with the use of sliding glass doors opening to deck and patio. The large foyer offers access to all areas of the home.

First floor—2,086 sq. ft., Basement—2,086 sq. ft. Garage—831 sq. ft.

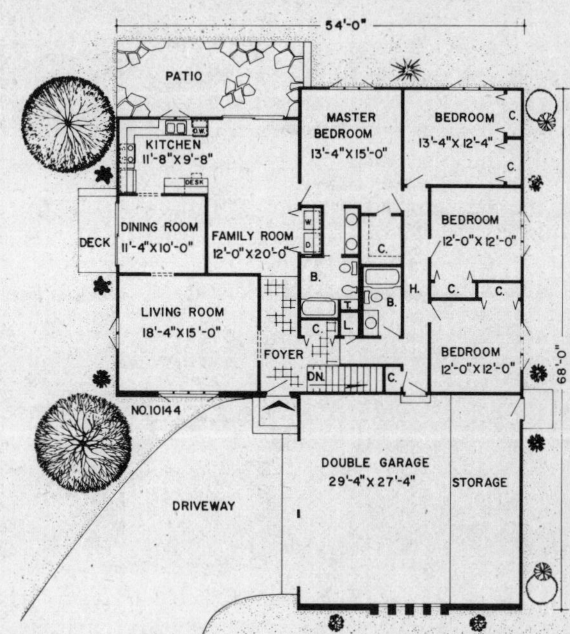

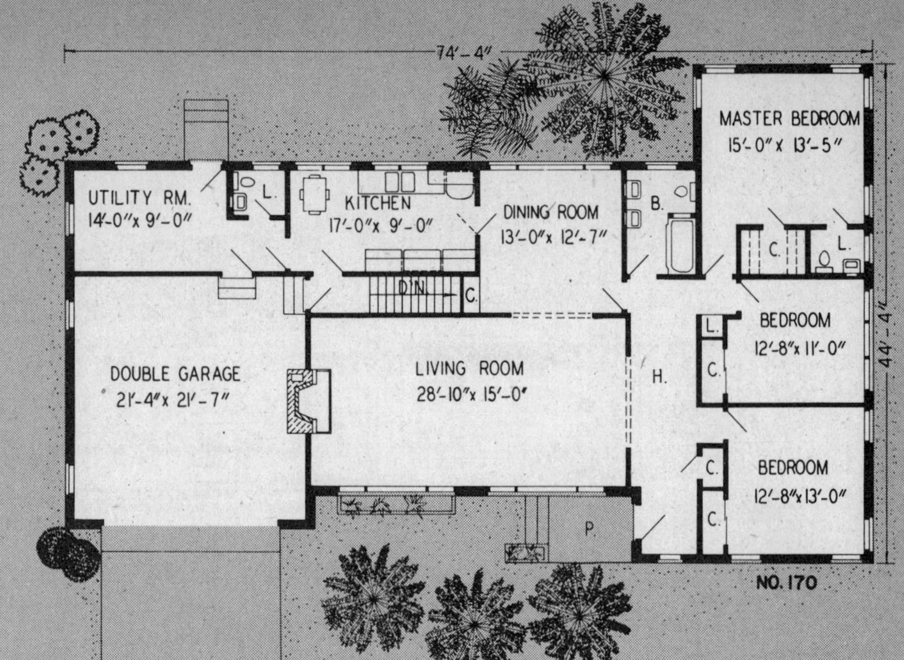

Cut stone veneer attractive

No. 170—The three bedrooms are large with many windows to assure adequate light and proper ventilation. A llarge walk-in closet and lavatory are shown in the master bedroom. The kitchen is large enough for a breakfast nook as shown and a utility room is provided. If a basement is not desired, the space allotted for the stairs can be used in the kitchen.

First floor—2,064 sq. ft., Basement—2,064 sq. ft. Garage—500 sq. ft.

Roof combination yields unique facade

No. 10030—Combining gable and mansard roof styles accented by a stone chimney produces a memorable facade in this sprawling design. Living room is singled out for the gable roof and also merits cathedral ceilings, gable end windows, exposed beams and wood-burning fireplace. Dining room is separate and augments kitchen eating space, and cozy family room opens to kitchen and terrace. Luxury is expressed in the master bedroom, which features compartmented bath and huge walk-in closet.

First floor—2,090 sq. ft., Basement—2,090 sq. ft. Garage—528 sq. ft.

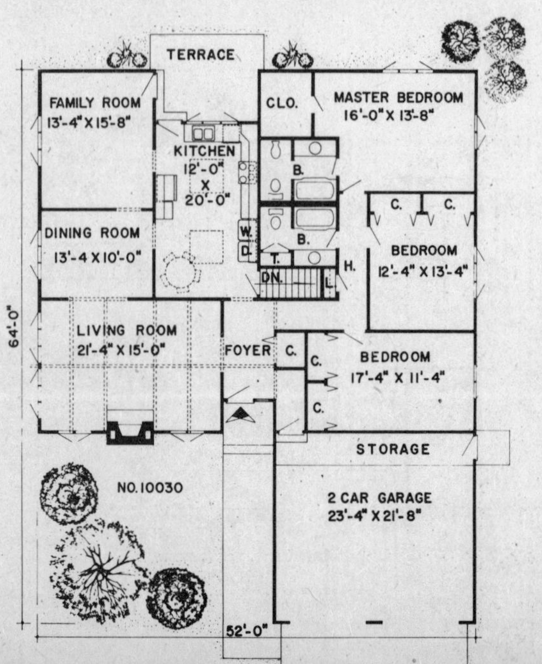

For price and order information see pages 108-109.

Triple-zoned home invites living

No. 10076—Expansive living areas and a luxury bed-room wing characterize this rambling Mansard roof plan. Exterior blends rough cedar plywood siding, vertical brick, and shake shingles, while the interior zones areas for sleeping, formal entertaining, and casual living. Opulent master bedroom suite boasts two huge walk-in closets, a sitting room and full compartmented bath. Another compartmented bath serves three sizable bedrooms, each replete with closet space. Kitchen, dining room, family room, and sitting room open to patio via sliding glass doors.

**First floor—2,672 sq. ft., Basement—2,672 sq. ft.
Garage—576 sq. ft.**

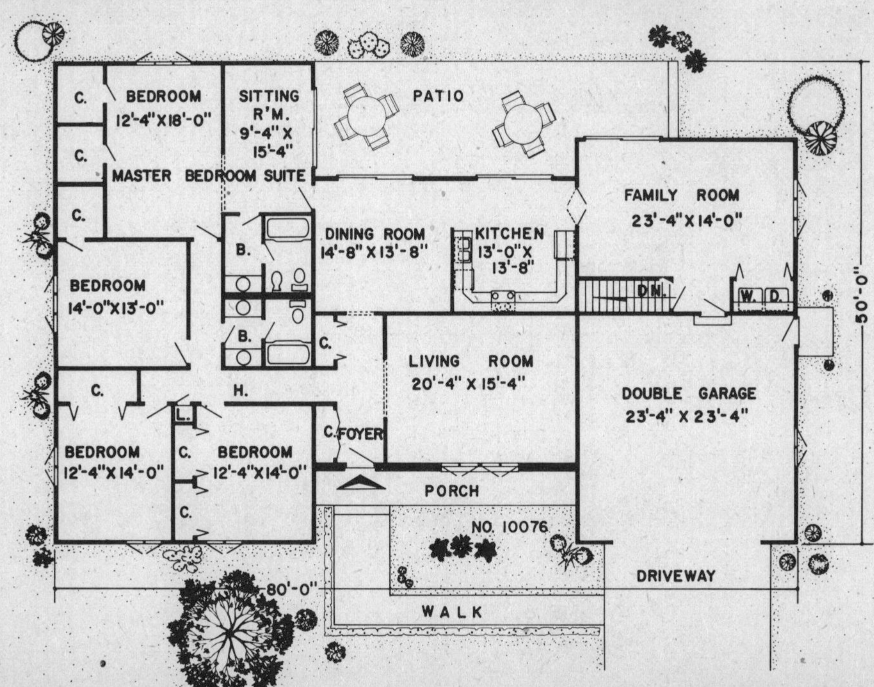

Patio contains built-in barbecue

No. 10064—Shake shingles, battened siding and natural stone are skillfully blended together to produce a beautiful facade for this home. The traffic pattern permits access to all rooms from the foyer without crossing another room. Sliding glass doors in the family room, dinette and master bedroom open onto the large patio-pool area. The built-in barbecue on the patio is roofed to provide a shaded area near the pool.

First floor—2,585 sq. ft., Basement—2,585 sq. ft. Garage—493 sq. ft.

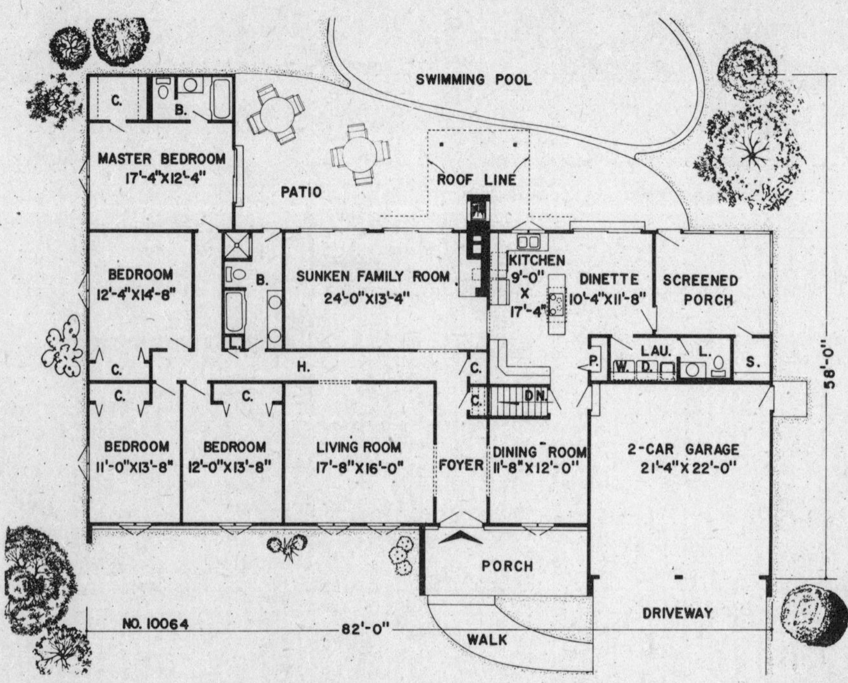

For price and order information see pages 108-109.

Great detail

No. 10032—A French mansard roof is combined with contemporary styling to produce this beautiful four bedroom home. Low maintenance materials such as shake shingles and rough cedar siding are used on the exterior. The master bedroom is extra large. It has a dressing area, two walk-in closets and a private, compartmented four piece bath. Sliding glass doors in the living room and bedroom open onto a private patio. The family room has a wood-burning fireplace. The busy housewife will enjoy the kitchen and adjoining laundry room. The large double garage (24' x 24') can have the doors installed at the front, side or rear, depending on the building site.

First floor—2,422 sq. ft., Basement—2,422 sq. ft. Garage—576 sq. ft.

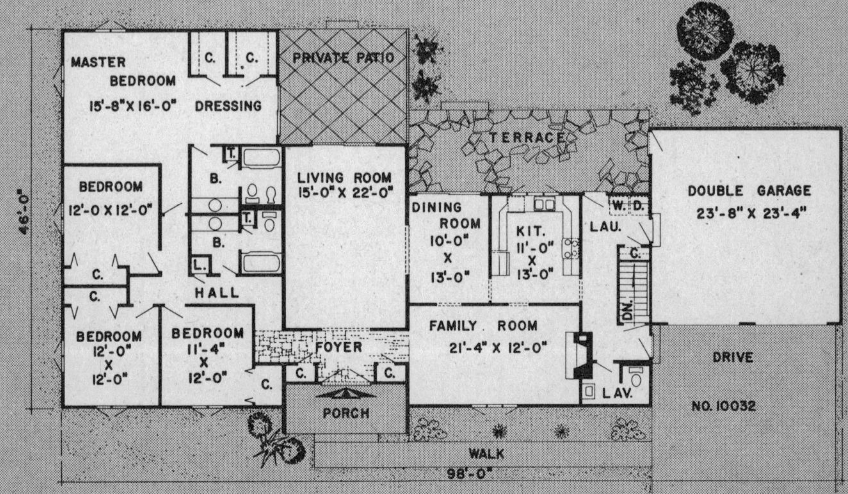

Modern convenience for young family

No. 354—Opening off a large foyer, the living room is well separated from the remainder of this home, allowing adults to entertain without being disturbed by children's activities. The kitchen/family room area is divided by a breakfast bar efficiently designed for easy living. Cathedral ceilings are used throughout to provide the extra atmosphere today's young families enjoy.

First floor—1,738 sq. ft., Garage—504 sq. ft.

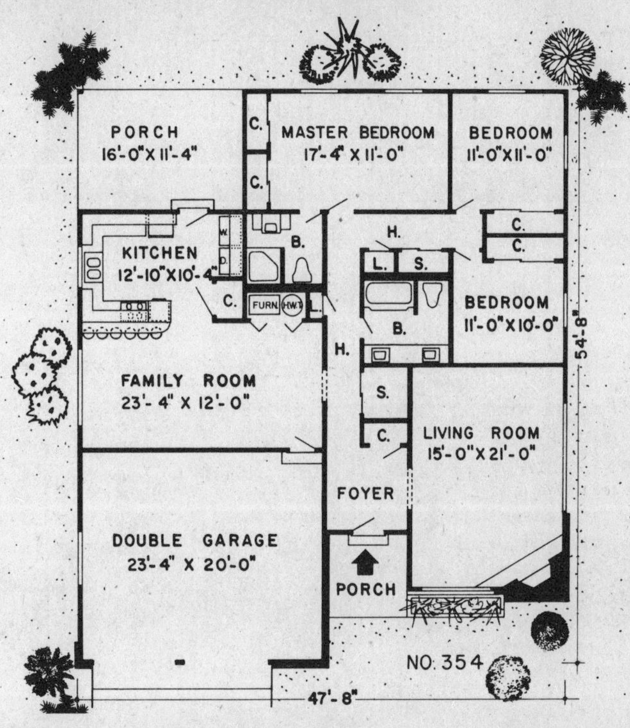

Fireplace inspires romantic dining

No. 9908—Pleasurable dining in the expansive living-dining area is created by the atmospheric wood-burning fireplace in this brick-layered traditional. A functional breakfast bar joins the kitchen and family room, which is placed to enjoy the terrace. A gracious foyer eliminates cross-traffic and allows access to the living or sleeping wing, where three sizable bedrooms and two full baths are provided. The double garage also opens to the terrace.

First floor—1,896 sq. ft., Basement—1,896 sq. ft. Garage—509 sq. ft.

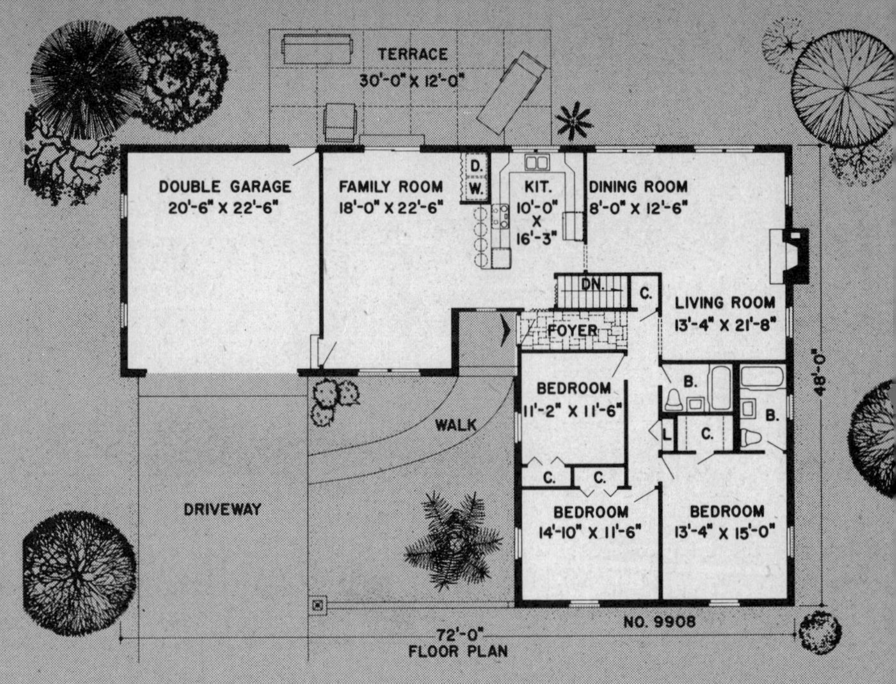

Nature emphasized in rambling plan

No. 9924—Rough-hewn Redwood siding and brick exterior fuses with water fountain-accented atrium to concoct a design that blends with nature and stresses interior livability. Closet-lined foyer steers traffic to formal living room with fireplace and terrace as well as to informal living areas. Kitchen and breakfast nook, separated by snack bar, open to another terrace, while the elongated family room merits sliding glass doors to the atrium. Four bedrooms, including master bedroom with dressing area and compartmented bath, comprise the sleeping wing, which attaches a laundry niche for convenience.

First floor—2,533 sq. ft., Basement—2,533 sq. ft. Garage—559 sq. ft.

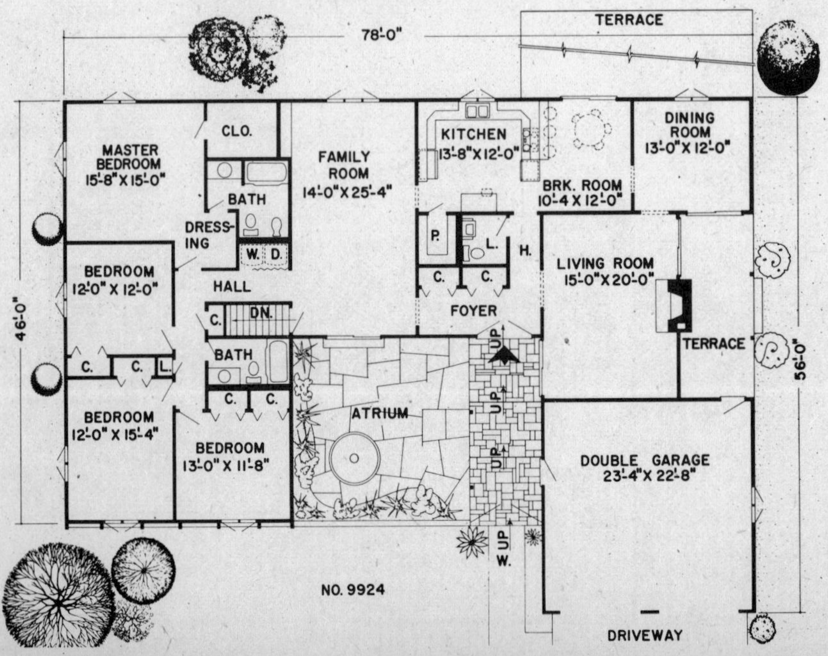

For price and order information see pages 108-109.

Hexagonal living room elevates design

No. 9954—Soaring above the main body of the home, the hexagonal living room captures the startling uniqueness of this design. The sunken living room comes alive with firelight and clerestory windows, and with the formal dining room, enjoys sliding glass doors to the terrace. The sizable kitchen apportions laundry and eating space, and adjoins the family room. Four ample bedrooms and two full baths comprise the sleeping wing, which includes a family bath with tub and shower stall.

First floor—2,210 sq. ft., Basement—754 sq. ft. Garage—576 sq. ft.

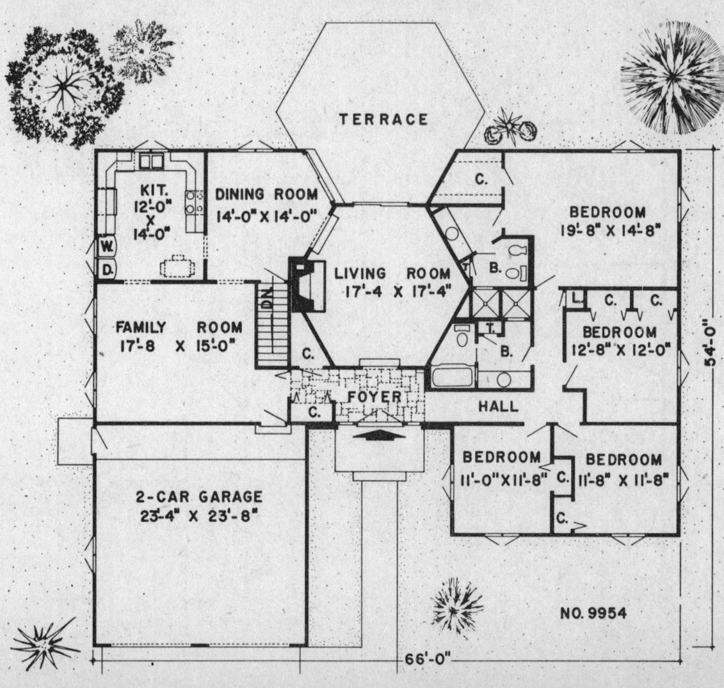

Deck frames master bedroom

No. 1060—Edging the master bedroom and immense living room, double triangular decks emphasize outdoor living in this creative three bedroom home. All areas of the home are heavily glassed, with windows and sliding glass doors used liberally, and have easy access to the outdoors. The living room extends over 39 feet and spotlights a central wood-burning fireplace. Other notable features of the home include a walk-in closet off the dining area, a large pantry, and a laundry room that borders the kitchen.

First floor—2,160 sq. ft., Basement—1,600 sq. ft. Carport—440 sq. ft.

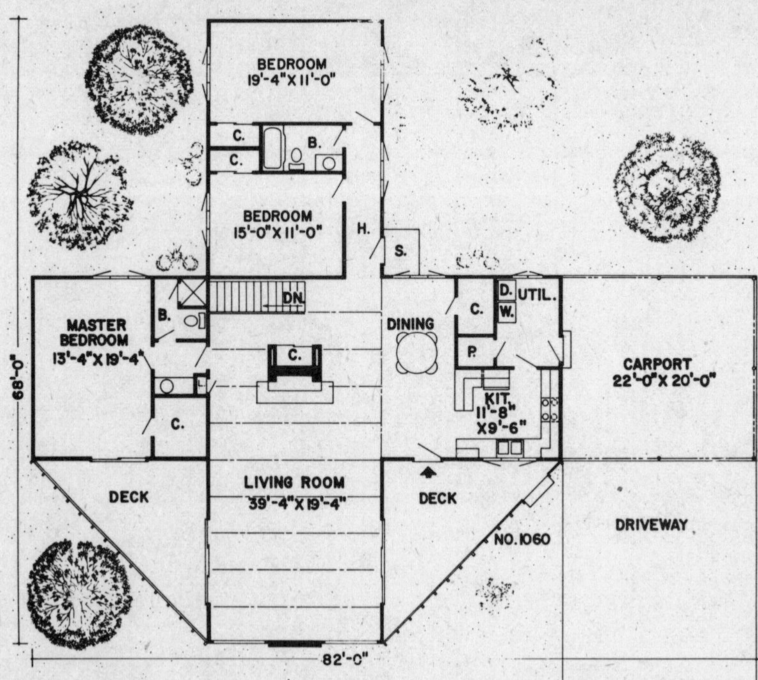

*For price and order information
see pages 108-109.*

Atrium highlights open design

No. 1054—Simple lines and spacious rooms mark this three bedroom contemporary which spotlights a central atrium for added light. Linked to the family room via sliding glass doors, the atrium also shares its bright, airy character with the living room dining room, and bedroom hallway. The family room boasts a raised hearth wood-burning fireplace, built-in cabinets, sliding glass doors to the patio, and an open connection with the kitchen. Formal living and dining rooms are featured, and three bedrooms sandwich two full baths.

First floor—2,368 sq. ft., Carport—400 sq. ft. Atrium—234 sq. ft., Outside storage—48 sq. ft.

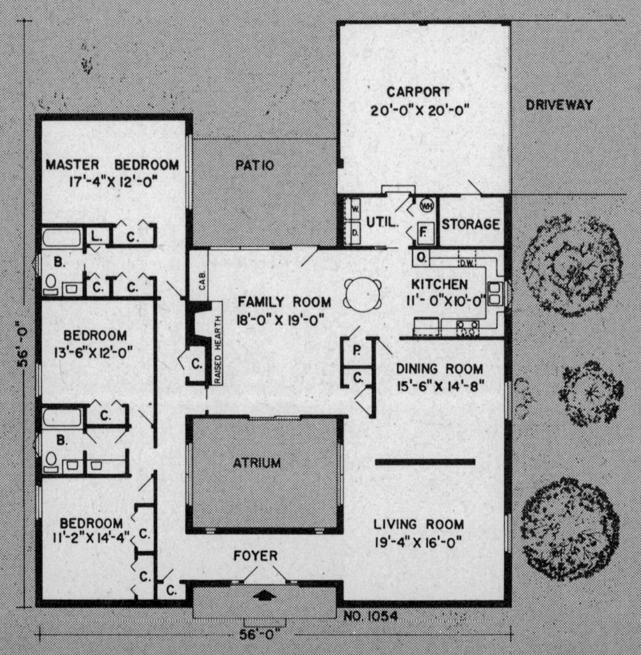

Cupola crowns well-placed garage

No. 9866—Trimmed with shutters, diamond light windows, and a cupola, the garage of this charming traditional exemplifies the successful fusion of appeal and convenience. The garage opens to the combination laundry and mud room opposite the kitchen. A spacious firelit family room opens to the terrace, where a built-in barbecue grill simplifies outdoor cooking. Three sizable bedrooms flank the formal living room and include a master bedroom with bath and double closets. A covered porch shades the front entrance.

First floor—1,768 sq. ft., Basement—1,768 sq. ft. Garage—528 sq. ft.

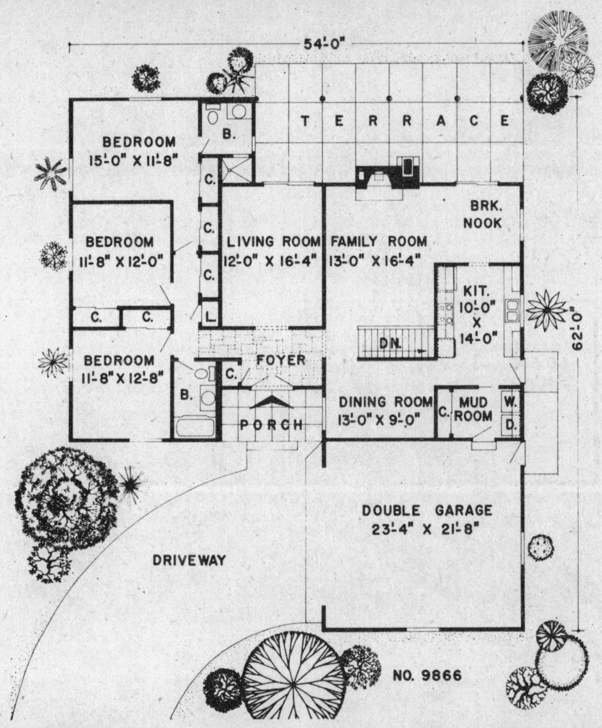

Stone exterior requires little care

No. 9874—Natural materials are the key to this four
bedroom home. The stone veneer requires little maintenance, the lovely stone terrace grants a remarkable
view, and the wood-burning fireplace cuts the chill on
winter evenings. Two full baths and walk-in closets,
plus a lounge area in the master bedroom, add to the
luxurious air of the sleeping area. The kitchen opens
onto a mudroom entering the garage. The center hall
plan preserves privacy and directs traffic to living and
bedroom areas.

**First floor—2,244 sq. ft., Basement—2,244 sq. ft.
Garage—591 sq. ft.**

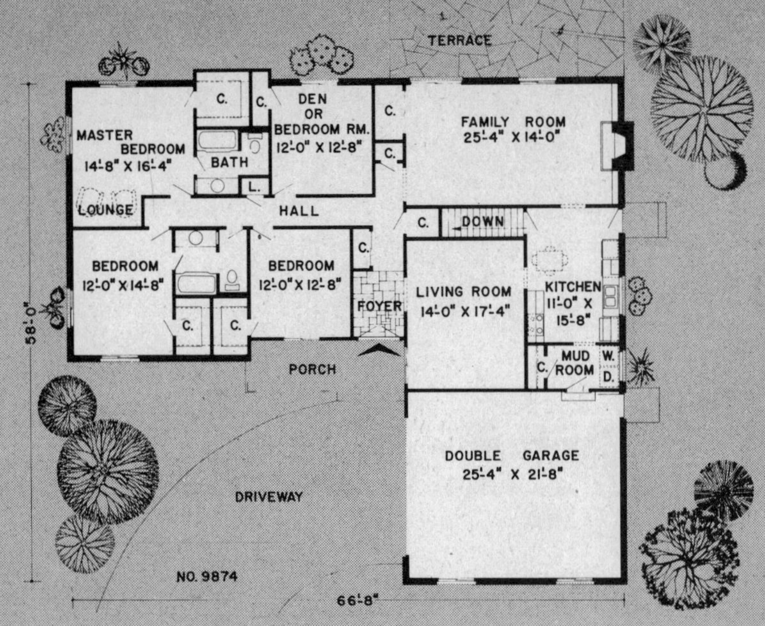

Dressing room favors master bedroom

No. 9911—Clean traditional lines, brick, and diamond light windows encase a design that specializes
in modern comforts and includes a master bedroom
with double closets, bath, and dressing room. In all,
four generous bedrooms and two full baths comprise
the sleeping quarters. Kitchen and family room work
together, divided by breakfast bar, and adjoining dining room opens to terrace. Wood-burning fireplaces
glow in both family room and living room, arranged
for quiet and formality.

**First floor—2,044 sq. ft., Basement—2,044 sq. ft.
Garage—572 sq. ft.**

*For price and order information
see pages 108-109.*

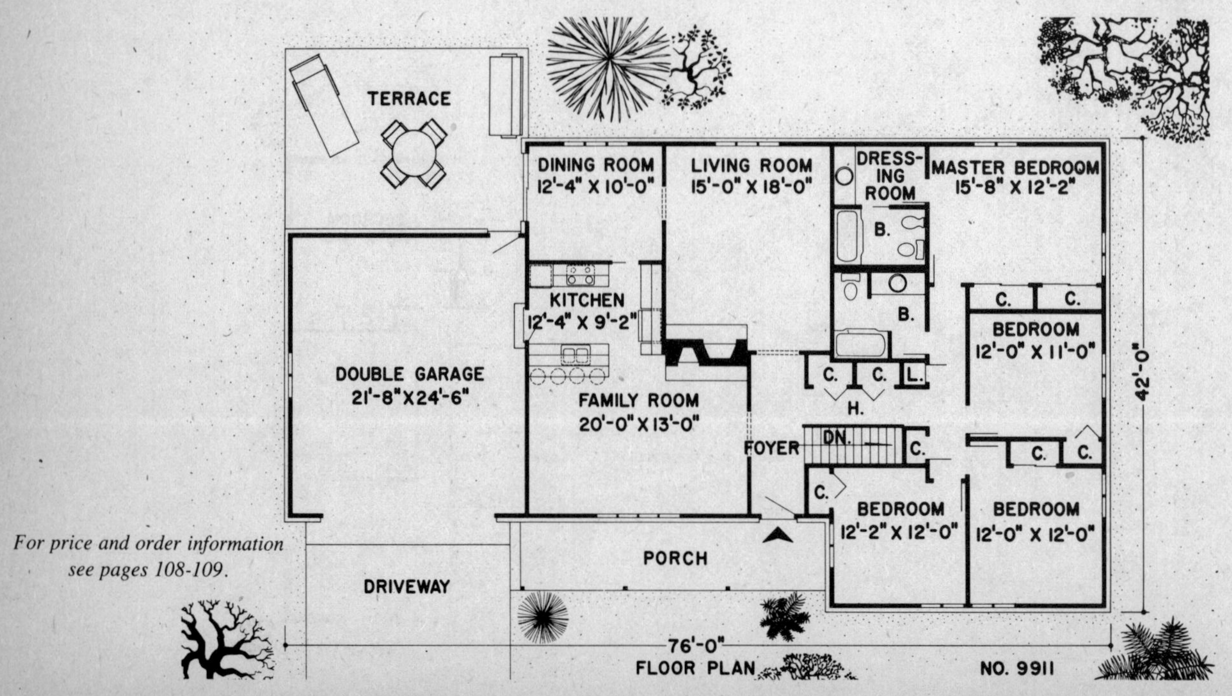

Fountain is a highlight

No. 9922—Imagine a colorful water fountain with automatic changing colored lights located on your front terrace. It will be enjoyed by passers-by as well as people seated in the living room. This is only one of many desirable features found in this beautiful four bedroom home. The large well planned kitchen is located between the formal dining room and the family room. The family room has a wood burning firpelace and a laundry niche for the washer and dryer. The master bedroom has its own private bath and three large closets.

First floor—2,261 sq. ft., Basement—2,261 sq. ft. Garage—535 sq. ft.

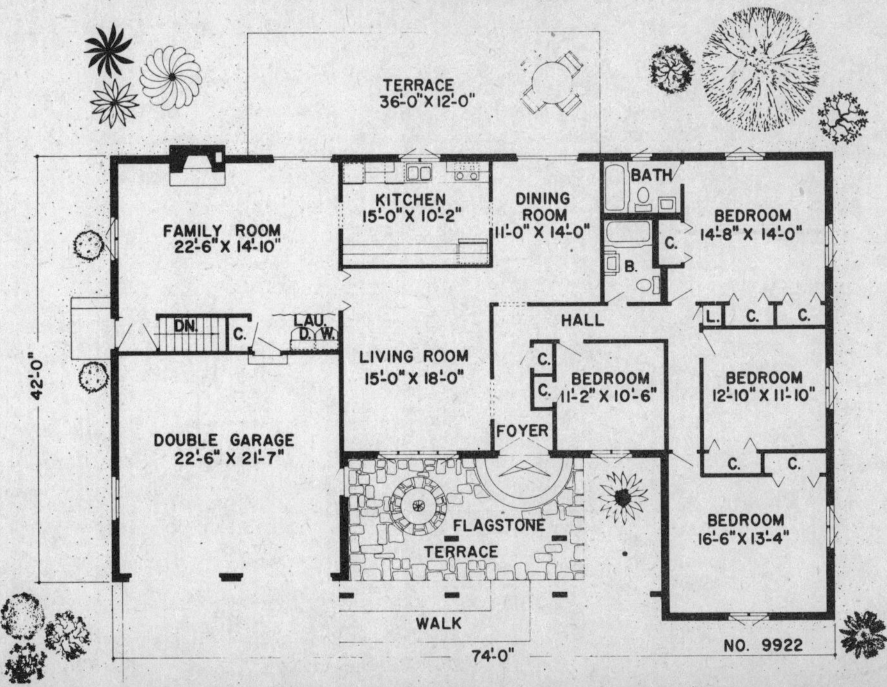

Master bedroom suite — luxurious and functional

No. 9966—Spanish styling is combined with a very desirable but practical floor plan arrangement. The sunken living room has a wood burning fireplace. The formal dining room is separated from the living room by an ornamental iron railing. The family room also has a wood burning fireplace as well as sliding glass doors which open onto the large terrace. Study the elegant master bedroom suite. It is a private world for parents, complete with a sitting room, compartmented bath, two large closets and a large bedroom area. Also, there are sliding glass doors which open onto the terrace.

First floor—2,396 sq. ft., Basement—2,396 sq. ft. Garage—576 sq. ft.

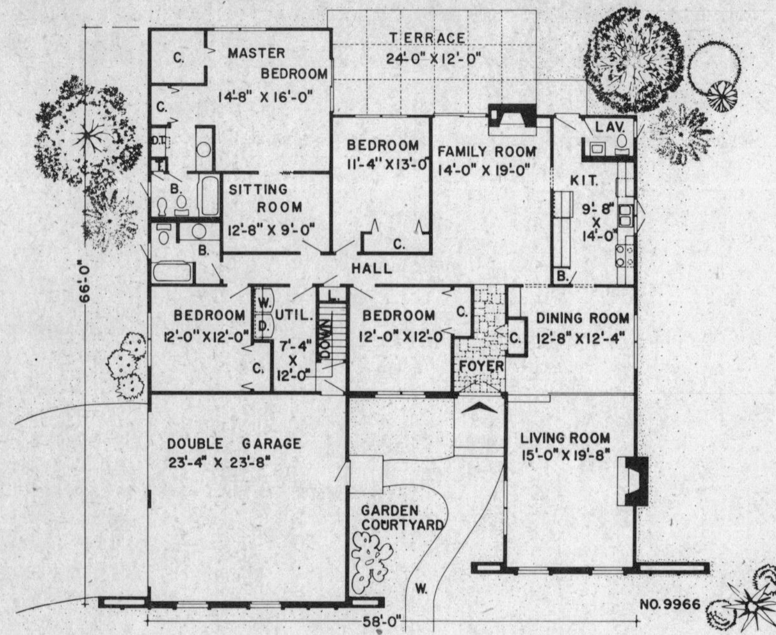

For price and order information see pages 108-109.

A home for relaxing

No. 9894—The brick veneer and rough cedar columns plus a shake shingle roof provide a low maintenance exterior. This will leave more leisure time for using the swimming pool, and sun bathing on the terrace. The master bedroom suite has its own private bath, walk-in closet, dressing table and sliding glass doors which look out over the pool area. The main bedroom wing has three large bedrooms and two full baths. The large family room can be closed off when desired.

First floor—2,327 sq. ft., Basement—2,327 sq. ft. Garage—534 sq. ft.

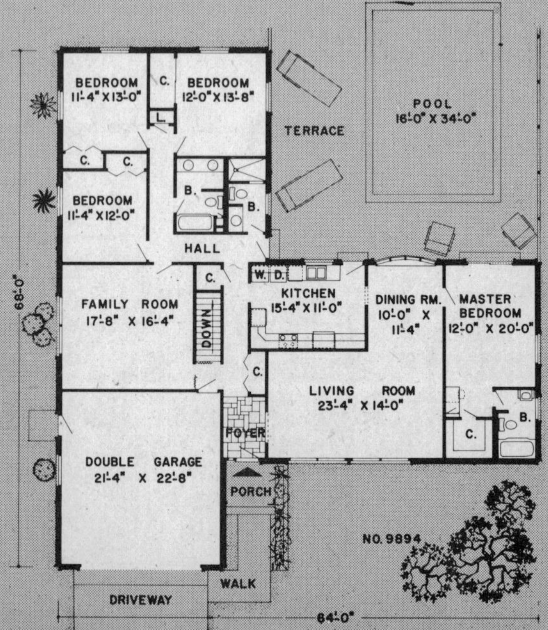

Semicircular terrace offers access

No. 9882—Spanning four rooms to the rear of the home, the semi-circular terrace in this plan is accessible through sliding glass doors from the living room, dining room and family room. The sunken living room with fireplace borders the formal dining room, and a kitchen with laundry space is situated to serve both dining room and family room. Three of the bedrooms, including the master bedroom which merits a bath and large closet, face front and enjoy lovely bay windows.

First floor—2,212 sq. ft., Basement—2,212 sq. ft. Garage—491 sq. ft.

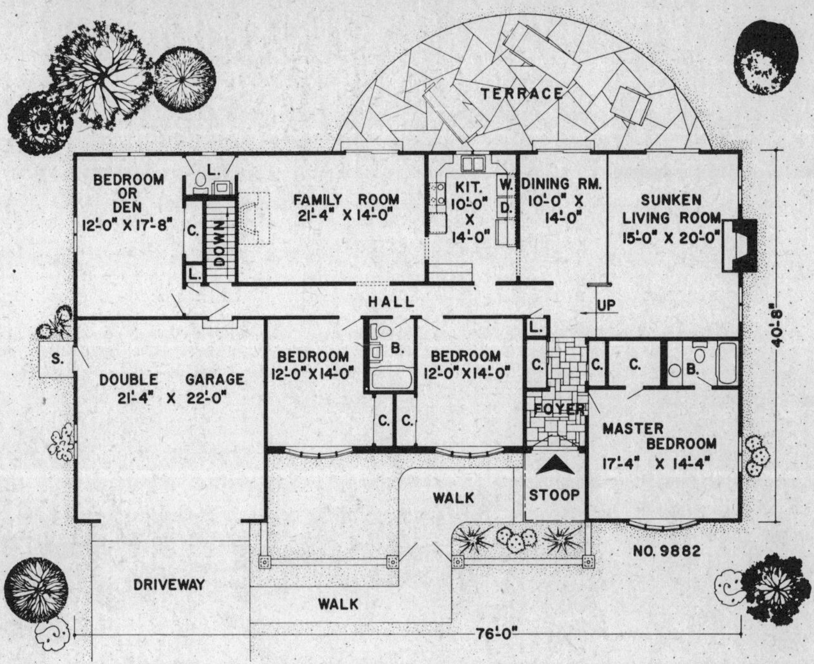

Stone, brick, glass light exterior

No. 150—Expanses of glass edge and encircle this L-shaped design and fuse with brick and stone to produce a unique and low maintenance exterior. Country kitchen joins the informal, enjoy able dining area, allots snack bar and laundry space, and serves living room and family room with equal ease. Sliding glass doors open family room to flagstone terrace, while the living room exhibits a cozy corner fireplace. Two large tiled baths, one private to the master bedroom, serve the sleeping area.

First floor—1,864 sq. ft., Basement—1,864 sq. ft. Garage—528 sq. ft.

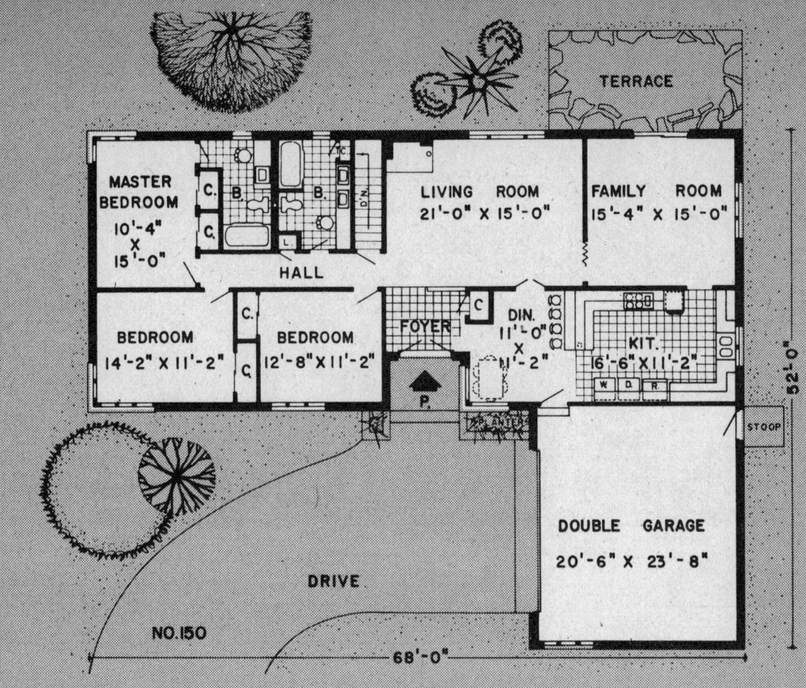

Angled design yields interesting shapes

No. 140—Fresh, innovative planning gives birth to a fascinating and unconventionally shaped kitchen, family room, and dining room in this contemporary design. This unique room arrangement also enjoys sliding glass doors to the partially roofed terrace, where a built-in barbecue grill invites outdoor cooking. The formal, fireplace-brightened living room is free of cross traffic. Three bedrooms comprise the sleeping wing, including a sizable master bedroom with private bath and double closets.

First floor—1,735 sq. ft., Basement—1,187 sq. ft. Garage—498 sq. ft.

For price and order information see pages 108-109.

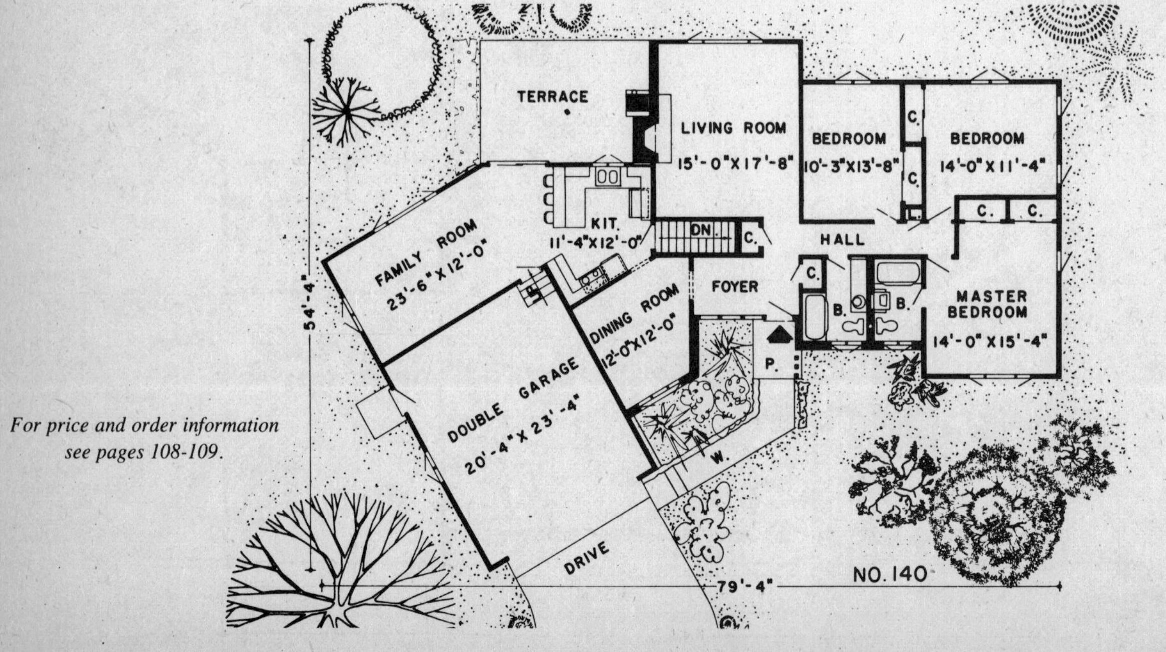

Laundry nook adds convenience

No. 9078—Opening from the kitchen and double garage is a large work room. Designed with a cut stone front, the home is ideal for entertaining. A foyer provides direct access to the family room, adjacent to the kitchen and rear terrace. Providing added convenience are a full basement and formal dining room. Three bedrooms, with two full baths, open onto a hall, hidden from main traffic areas of the house.

First floor—1,855 sq. ft., Basement—1,762 sq. ft. Garage and storage—565 sq. ft.

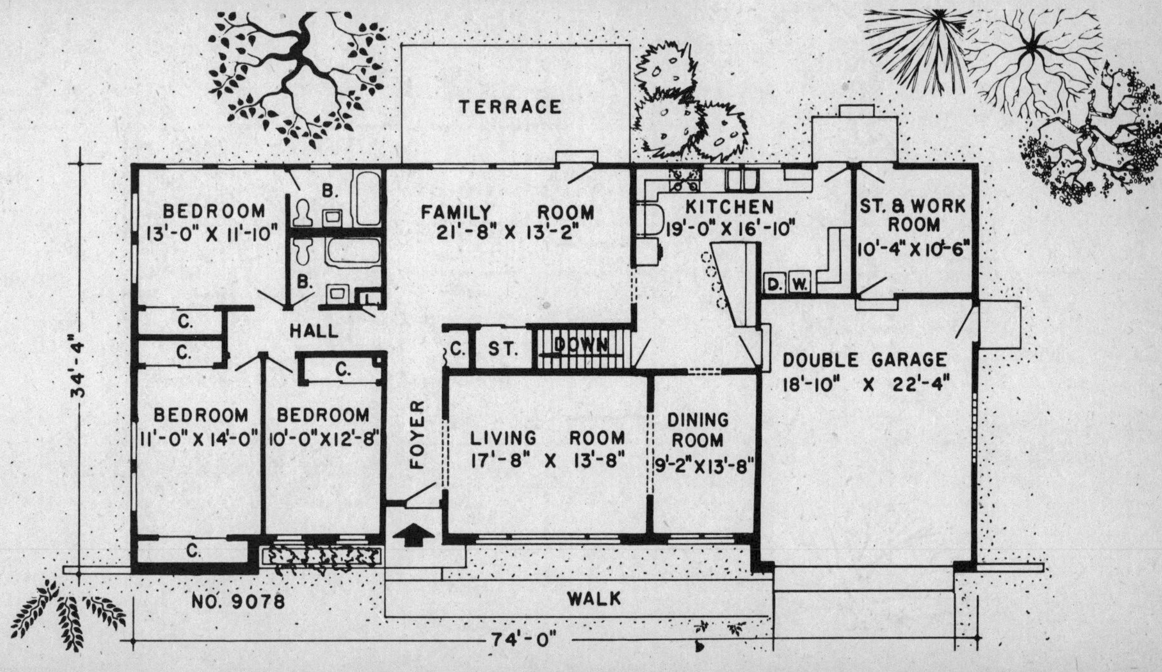

Kitchen complex unique, functional

No. 1002—Nothing succeeds like convenience in the cooking-dining-relaxing complex of this brick-sheathed three bedroom ranch style. Appendaging a den or family room, breakfast room, laundry niche and snack bar, the kitchen complex becomes a unique family center with access to front entry and carport. For guests, the adjoining dining and living rooms preserve formality and avoid cross-traffic. Three bedrooms include a 16-foot master bedroom, handsomely furnished with double closets and private bath. Two hall closets and a utility room serve the sleeping area.

First floor—1,765 sq. ft., Carport—360 sq. ft. Storage—144 sq. ft.

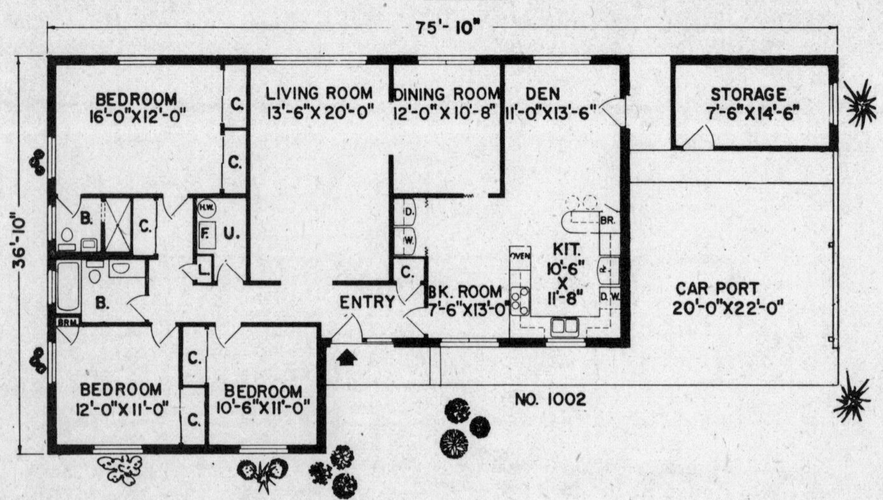

*For price and order information
see pages 108-109.*

Brick jackets living room wall

No. 318—Brick layers the entire wall on the fireplace end of the bow-windowed living room in this attractive home. Tiled foyer circulates traffic to living areas, sleeping areas, and basement. Divided by a long breakfast bar, kitchen and dining rooms are open and spacious, with dining room opening to terrace via sliding glass doors. Four bedrooms are indulged with closets and two tiled baths, while a den borders living areas and is adaptable to playroom or extra bedroom.

First floor—1,660 sq. ft., Basement—1,660 sq. ft. Garage and workshop—596 sq. ft.

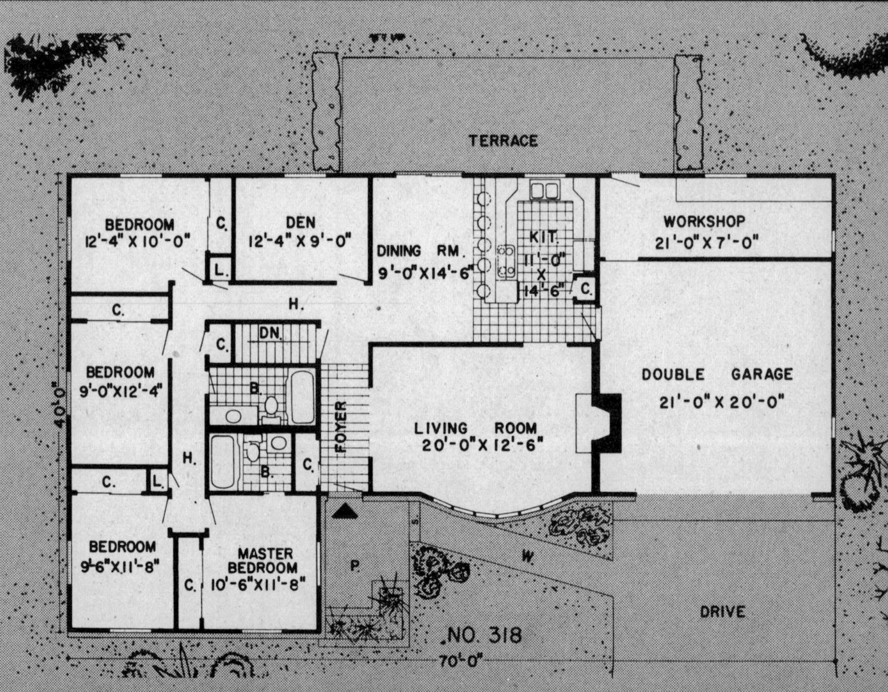

The ultimate to living

No. 254—This three bedroom Ranch style house was designed for an active family requiring extra large rooms. The oversized foyer adds an air of luxury and provides privacy for the living room. A separate dining room is provided so the family room can be devoted entirely to family living. The master bedroom has its own private bath and two large closets, both over five feet long. The kitchen contains the washer and dryer and a breakfast bar. A full basement is included for utilities, storage, etc.

First floor—1,900 sq. ft., Basement—1,900 sq. ft.

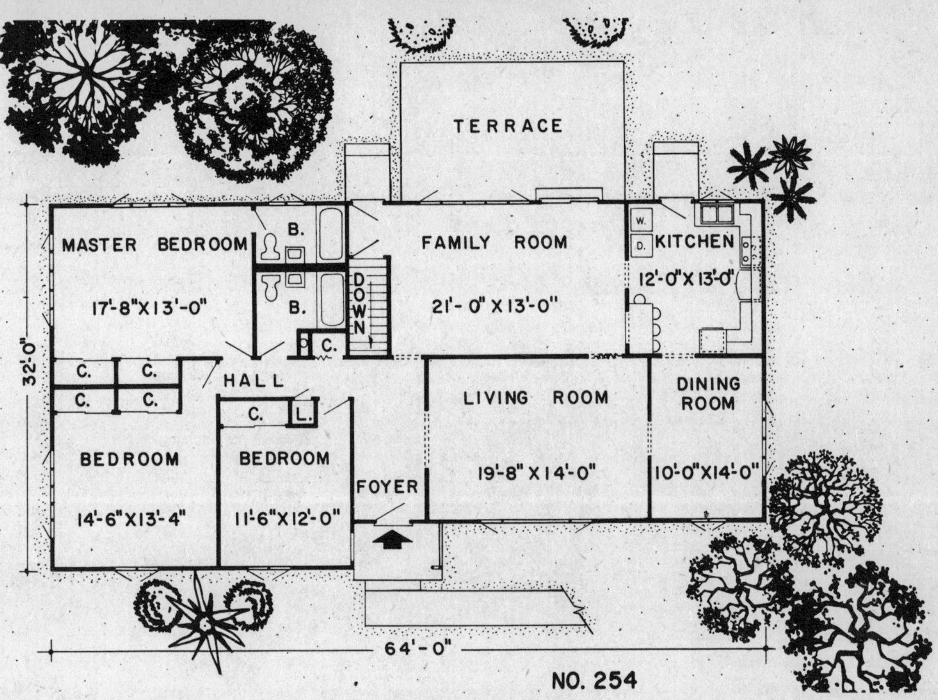

Barbecue encourages outdoor living

No. 348—Entrance to this home is gained through a large foyer opening to a central hall. The large versatile living room/dining room area is centrally located between the kitchen and family room for maximum usage. Incorporated into the kitchen, itself is a breakfast area suitable for most of a busy family's meals.

First floor—2,094 sq. ft., Garage—492 sq. ft.

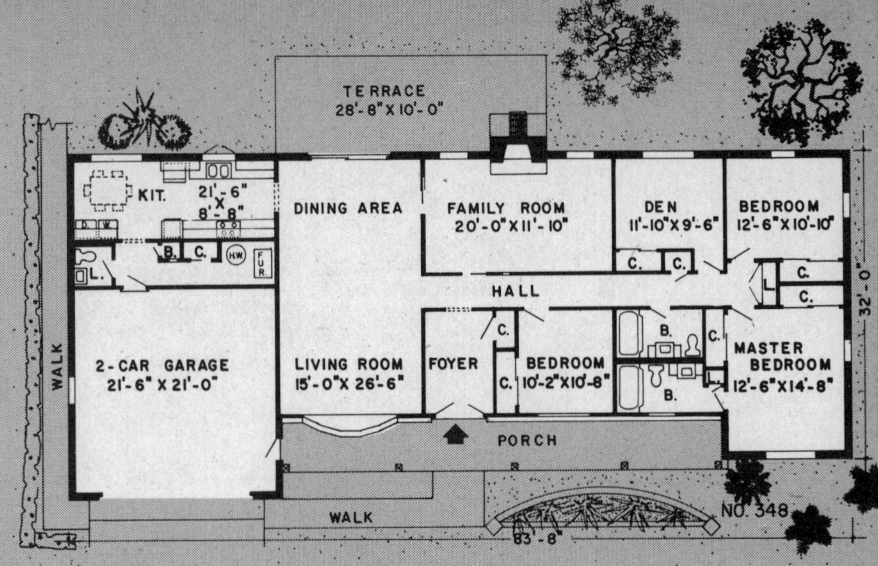

*For price and order information
see pages 108-109.*

From the family room, to the terrace . . . relax

No. 288—For the family who likes to entertain, the modern exterior of this home encloses a plan equally as modern. The center hall plan allows movements from the entrance to each of the primary home areas. For gracious entertaining the living room offers an attractive setting and for relaxed informality there is a large family room with a wood-burning fireplace and an attractive snack bar. The half bath is conveniently located near the center of the house.

First floor—1,775 sq. ft., Carport—372 sq. ft.

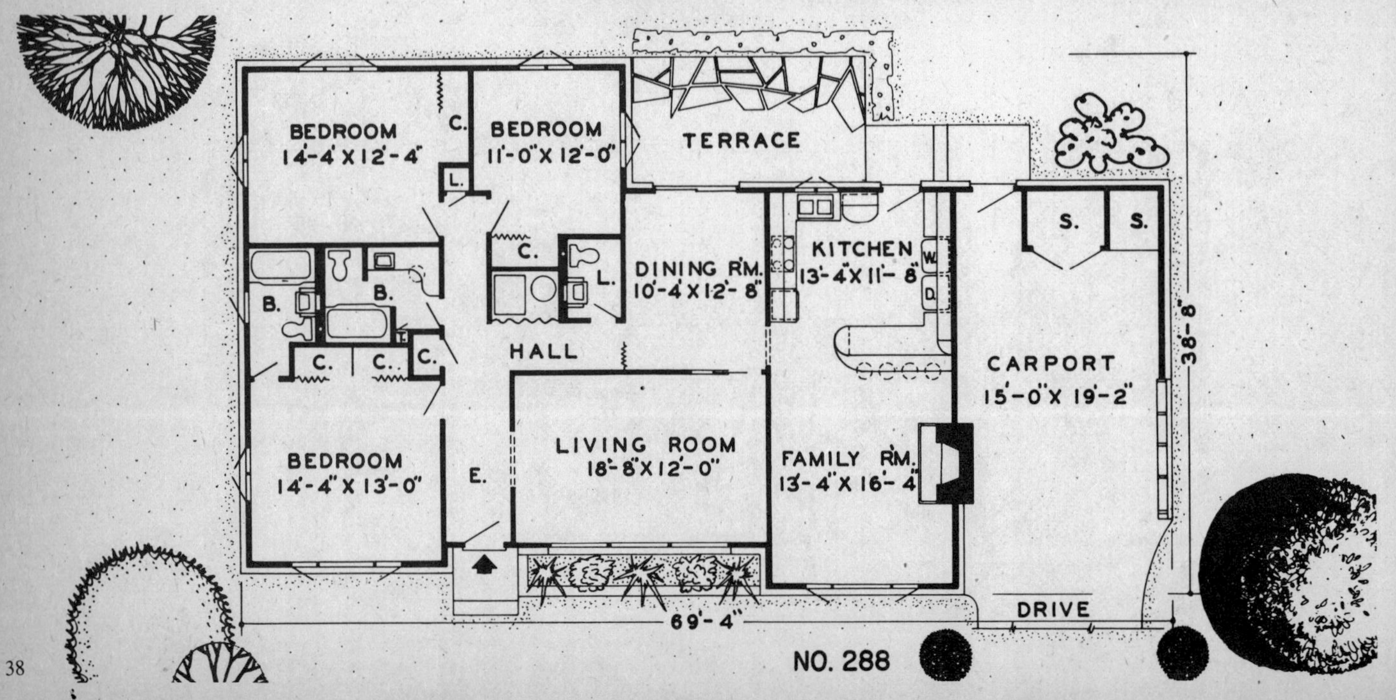

Bookshelves line living room

No. 308—Brightened by charming small-paned windows and meriting a television niche flanked by bookshelves, the living room of this comfortable plan offers a haven for quiet relaxation. Formal dining room enjoys access to living room and kitchen, and utility room, useful as a mud room, spans 19 feet to provide room for storage and laundry equipment. Two baths and plenty of closet space characterize the bedroom wing, master bath features shower stall.

First floor—1,845 sq. ft., Garage—436 sq. ft.

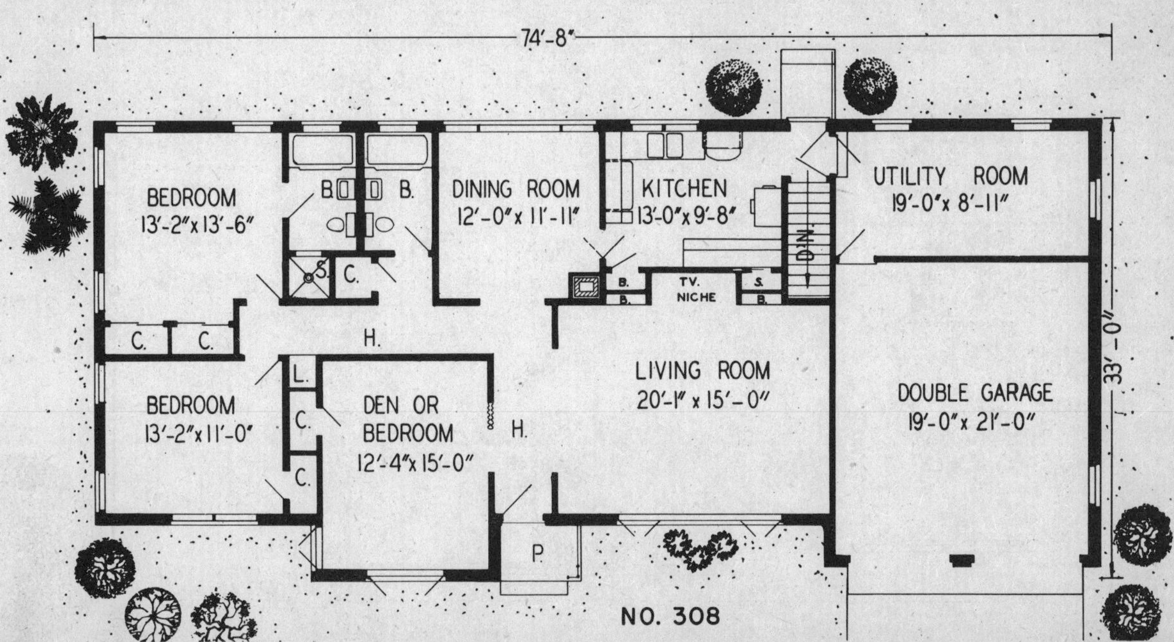

Old English brick graces exterior

No. 9221—Stately columns, Old English brick, and regal bay windows blend to form a magnificent exterior for this Colonial plan. Inside, the floor plan is efficient, beginning with the closeted foyer. Convenient to the kitchen, the formal bay-windowed living room boasts a wood-burning fireplace and is a natural setting for entertaining. Large family room with dining area is open to the kitchen on one side and to terrace via sliding glass doors. Three bedrooms include master bedroom with private bath and front bedrooms with bay window.

First floor—1,812 sq. ft., Basement—1,812 sq. ft. Garage—630 sq. ft.

For price and order information see pages 108-109.

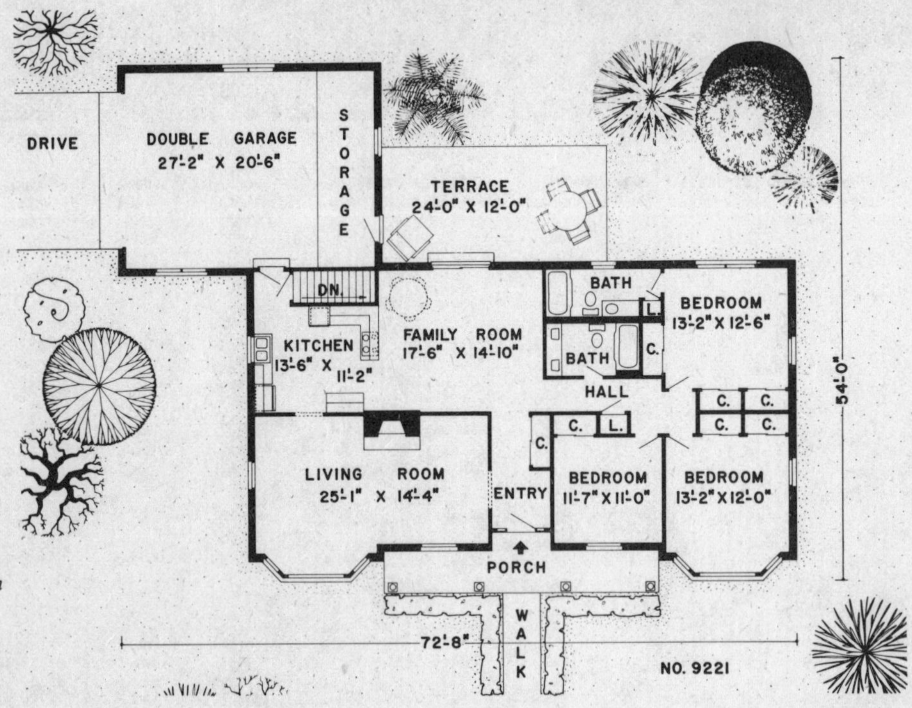

All space utilized

No. 9183—It is very apparent that this house was designed by a professional. The low silhouette and sharp distinct lines identifies it as a quality home throughout. An excellent floor plan arrangement utilizes all space to the maximum advantage. All rooms are large and there is plenty of closet and storage space. The family room features a built-in barbecue. It also has sliding glass doors which open onto the large flagstone terrace at the rear. A very useful mud room contains the laundry equipment and backs up to a bath which includes a shower. The garage entrance can be installed in any of three exposed sides.

First floor—2,198 sq. ft., Basement—2,030 sq. ft. Garage—572 sq. ft.

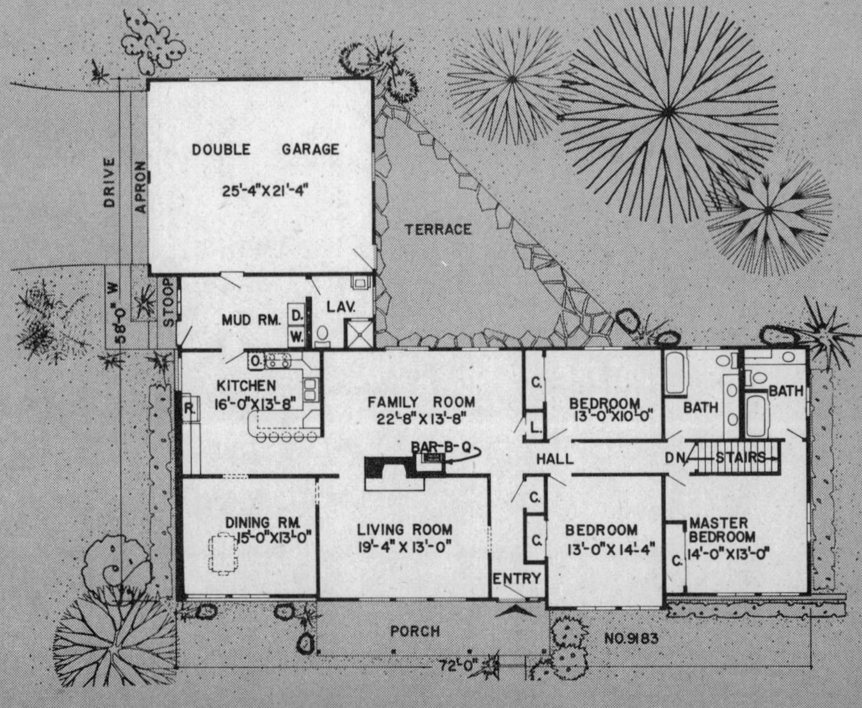

Three bedrooms two and one half baths

No. 9370—Centrally located, the living room and dining room of this three bedroom home gives the option of either daily use or being reserved for special occasions. Behind the master bedroom is a private garden accessible through sliding glass doors or from the garage. Because it is located in the rear, the garage can not detract from the street appearance of this home.

First floor—1,846 sq. ft., Basement—1,856 sq. ft. Garage—586 sq. ft.

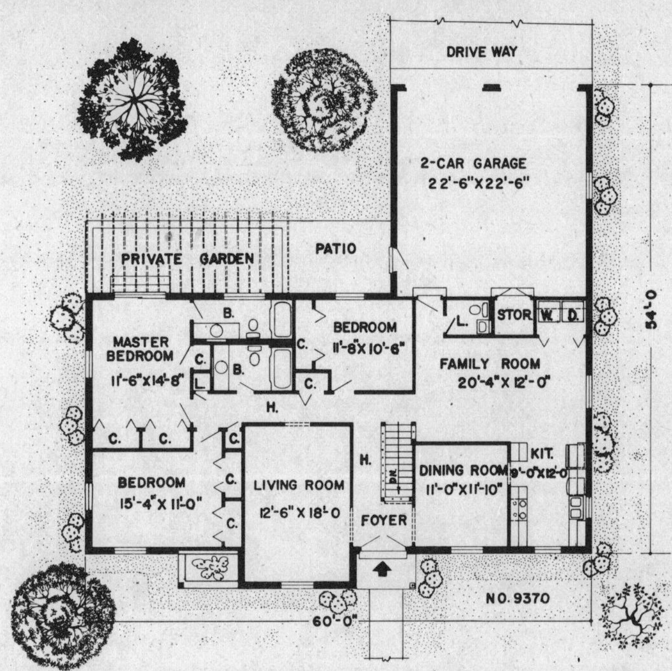

Brick, diamond windows blend perfectly

No. 9360—This well designed French Provincial design is a beautiful home. The brick, diamond windows, cupola, shutters, and color scheme all blend together perfectly. The floor plan is equally desirable. The three bedrooms are served by two full baths and there is a half-bath combined with the laundry, next to the family room. Sliding glass doors in the family room open onto a patio at the rear.

First floor—2,055 sq. ft., Basement—2,063 sq. ft. Garage—517 sq. ft.

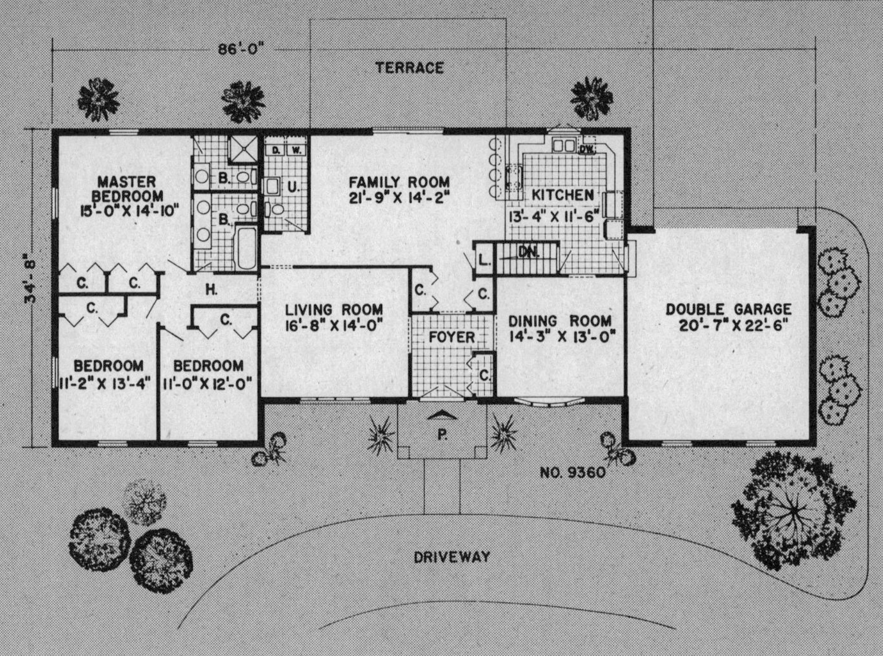

For price and order information see pages 108-109.

Stone, brick texture rustic facade

No. 9382—Battened plywood siding, fieldstone, and textured brick layer the facade of this roomy ranch style to create an exterior with rugged appeal. Entry is across the covered porch into a gracious foyer with living room at left and formal dining room at right to facilitate entertaining.

First floor—2,440 sq. ft., Basement—2,440 sq. ft. Garage—578 sq. ft.

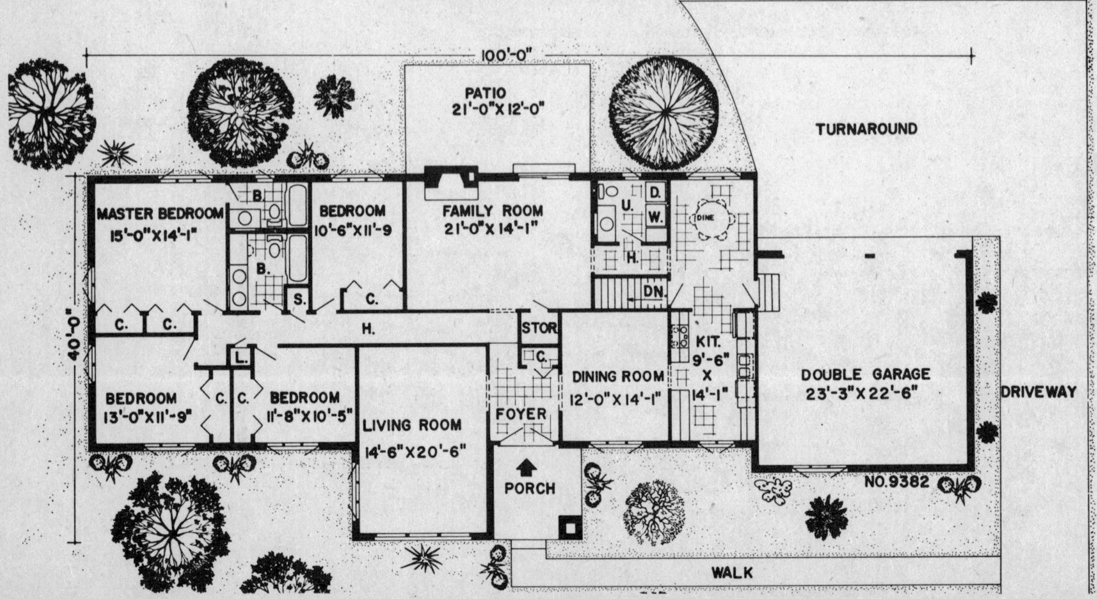

Privacy stressed in Contemporary

No. 9117—The living room and family room have been purposely separated, in this contemporary home, to provide maximum privacy to both rooms. This will permit activities of different noise levels to exist without disturbing people in the other room. The kitchen area will please the housewife. It contains a built-in range and oven, dishwasher, breakfast bar and an abundance of cabinets. Close by is the utility room which houses the washer, dryer and freezer. The storage room behind the utility room provides space for items used out of doors.

First floor—1,982 sq. ft.

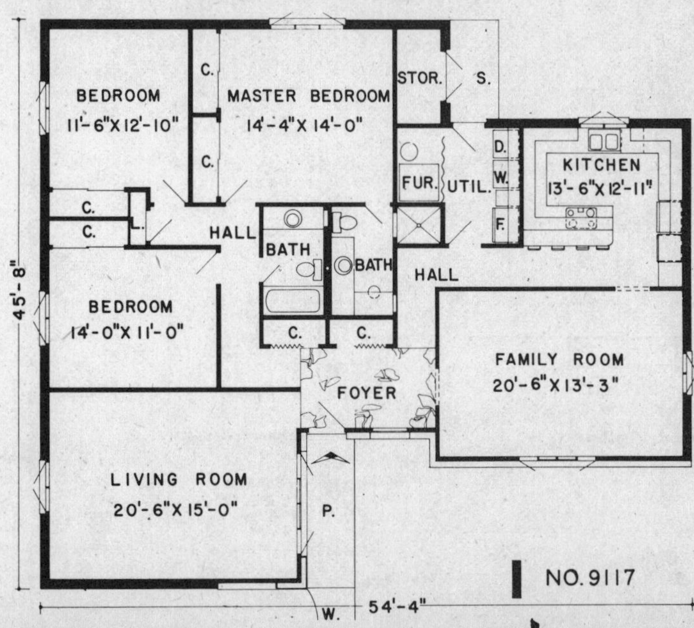

Different outside . . . fabulous inside

No. 10010—Shake shingles and battened rough cedar
plywood siding are specified for the exterior of this
Mansard roof design. These materials require a minimum of care and maintenance. The floor plan is excellent. The foyer leads to four bedrooms, and two full
baths. The living room and family room are both
large. Both open onto a large terrace. The kitchen
contains a planning desk, the laundry equipment, and
an abundance of cabinets. A double garage and full
basement also are included.

**First floor—2,153 sq. ft., Basement—2,153 sq. ft.
Garage—583 sq. ft.**

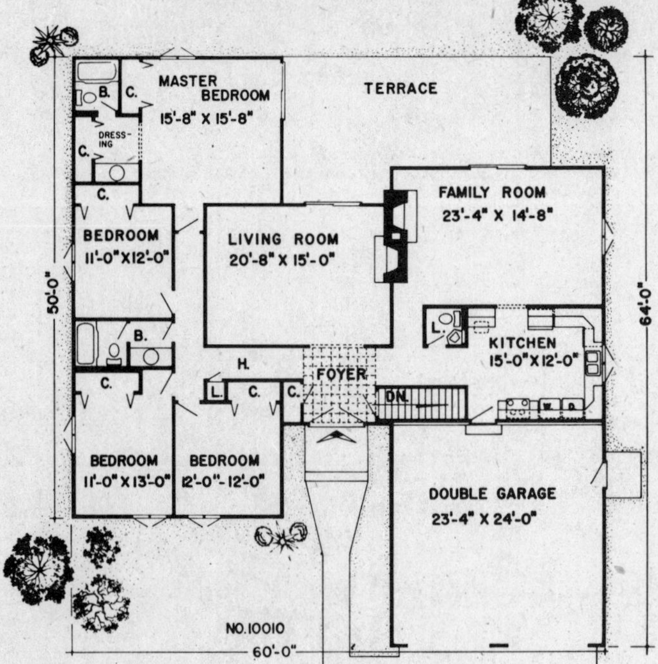

*For price and order information
see pages 108-109.*

Atrium illuminates entire home

No. 10028—Emerging as the central point of this
contemporary home, the atrium exhibits an ornamental water fountain and the atrium floods much of the
home with natural light. Walled mostly in glass, it
transmits daylight and starlight to the kitchen, dining
room, and the family room which is edged by a
terrace. A formal living room is isolated toward the
front of the design, as is the master bedroom suite
featuring bath, built-in vanity and walk-in closet.

**First floor—2,148 sq. ft., Atrium—380 sq. ft.
Basement—768 sq. ft., Garage—576 sq. ft.**

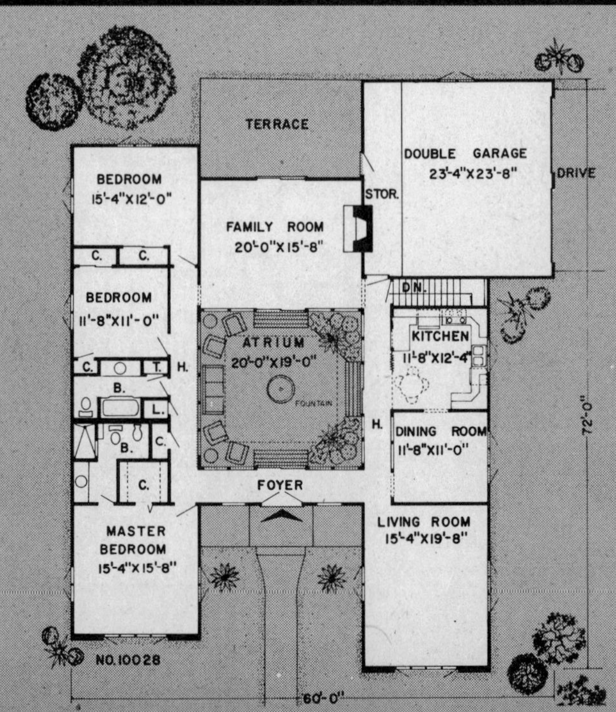

Simplicity sparks Spanish design

No. 10008—Fusing Spanish arches with a stark simplicity evokes a dramatic exterior that previews the impressive floor plan of this brick and stucco design. Entry is into the gracious foyer with an arresting view of the sunken living room's wood-burning fireplace. Exposed wood beams accent the ceiling of the room, adjacent to the firelit family room with access to the terrace. An extended kitchen offers a pantry and dining space and borders a convenient half bath. Four bedrooms are outlined, including a lavish master bedroom with closet-lined dressing area and compartmented bath.

First floor—2,672 sq. ft., Basement—2,672 sq. ft. Garage—576 sq. ft.

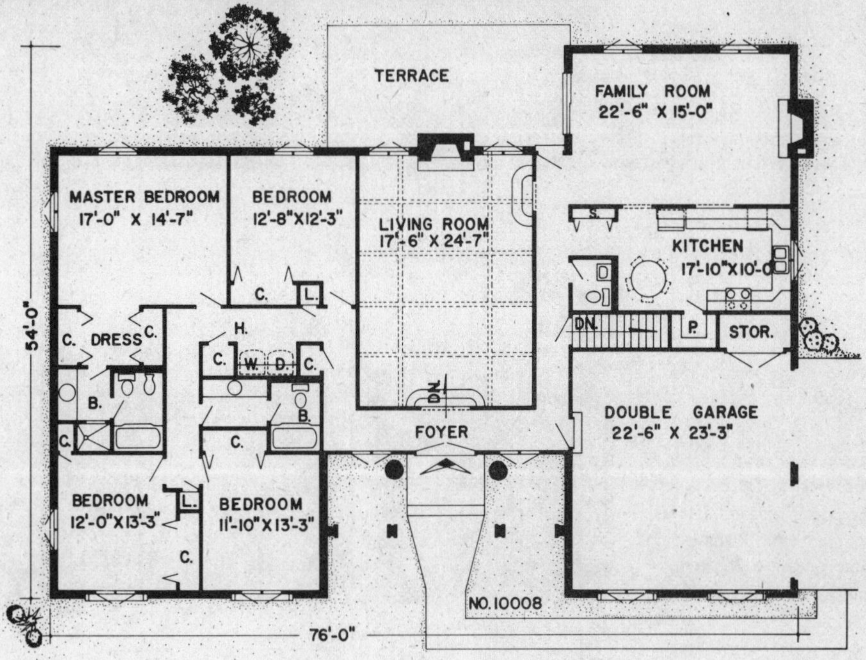

Spanish styling suits narrow lot

No. 10208—Elegant, elongated, and definitely Spanish in origin, this three bedroom home is accented with patios and designed to grace a narrow lot. Front entry garage borders an open patio with arched walkway which preludes entry into the home. Conveniently open to kitchen and family room, a covered patio is sandwiched between garage and living areas. Double closets line the foyer, which is flanked by living room and family room. Bedrooms are served by two full baths.

**First floor—1,600 sq. ft., Basement—1,600 sq. ft.
Garage—586 sq. ft.**

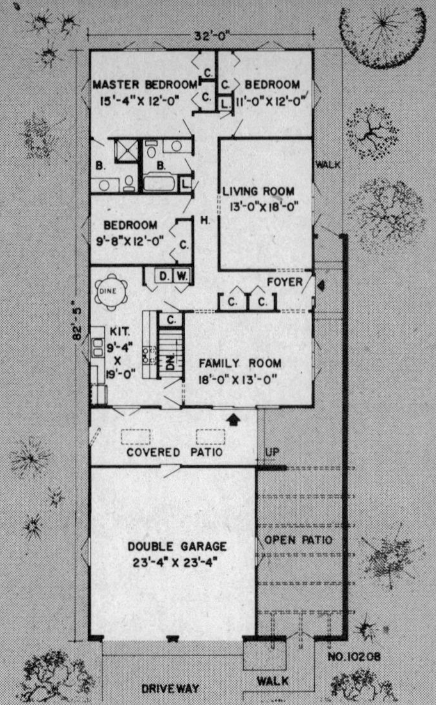

Two decks embrace nature

No. 10284—Built around and into its surroundings, this three bedroom design features an exterior that blends with nature and double decks to embrace it. Living and dining rooms encircle one deck, opening through three sets of sliding glass doors, while another deck opens exclusively to the master bedroom. Favored with walk-in closet and full bath, the master bedroom is set apart for privacy. Two other bedrooms, a hall bath and laundry room complete the plan, and a large detached double garage is provided.

First floor—1,568 sq. ft., Garage—484 sq. ft.

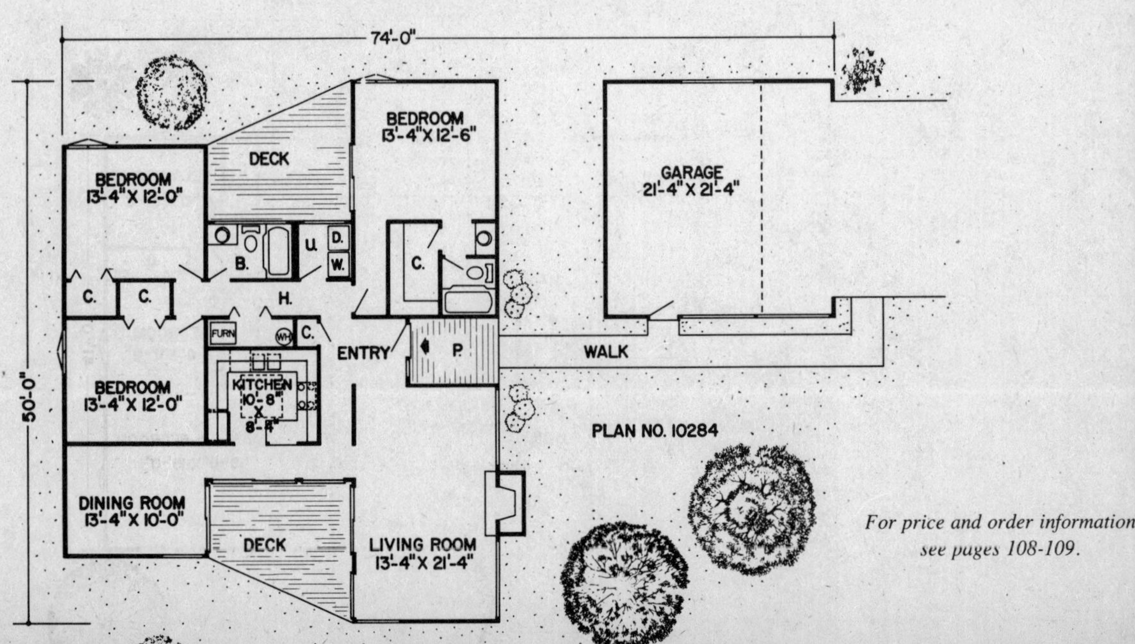

*For price and order information
see pages 108-109.*

Contemporary elects exceptional exterior

No. 10210—Two vaulted gambrel roofs are trimmed with a unique window treatment and cedar shakes to fashion an attractive double-winged contemporary design. The roofs camouflage the double garage and accent the bedroom wing, with the master bedroom treated to an escalating expanse of windows. The foyer invites entry to formal living room and provides access to informal kitchen, family room, and patio beyond. Combination laundry room and half bath border the family room, and the well-placed dining room will fill entertaining needs.

First floor—2,182 sq. ft., Basement—2,182 sq. ft. Garage—480 sq. ft.

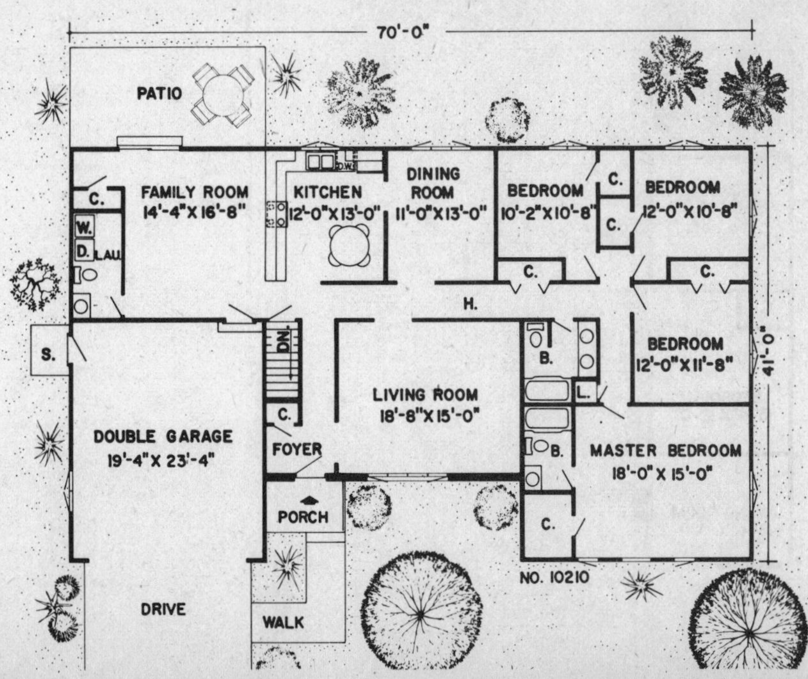

Charming exterior highlights design

No. 382—Attention to style in this contemporary begins with the exterior, with shutters, diamond light windows, and prominent stone chimney. Inside, the plan combines fashion and comfort, offering a crackling wood-burning fireplace in the living room, a closeted foyer with access to the garage, and a functional kitchen complex complete with snack bar. Three bedrooms are served by two full baths, and washer-dryer niche. Garage and dining area open to the semi-private patio at rear.

First floor—1,535 sq. ft., Garage—475 sq. ft.

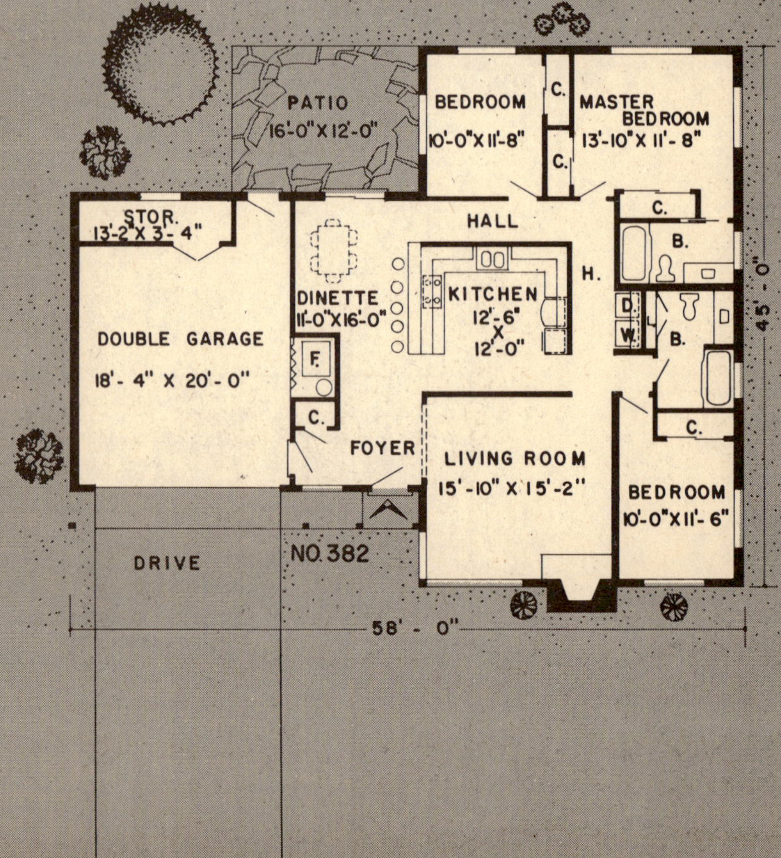

*For price and order information
see pages 108-109.*

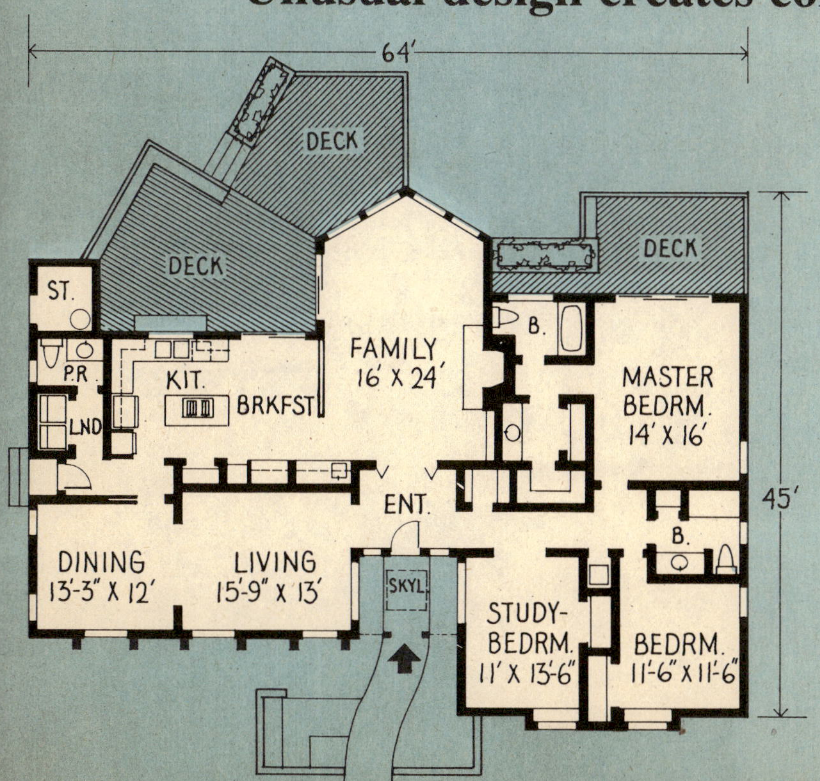

Unusual design creates comfortable living

No. 26760—The central focus of this highly pleasing 3 bedroom rancher is the family room, its largest most architecturally interesting space. The first room seen upon entering, this room features a prow shape, a beamed ceiling and a fireplace. Sliding glass doors give access to the multi-leveled deck. The well designed kitchen has a center work island and large breakfast area overlooking the deck. The dining room and living room are conveniently placed for ease of entertaining. The master bedroom has a private bath and dressing room. Also included are plenty of closets and a private deck. Two smaller bedrooms share a spacious bath.

**Living area—2,000 sq. ft.
(excluding decks)**

DESIGNER: Leo A. Kirkman

Charming 3 bedroom rancher

No. 26741—This charming home features something for everyone, the romantic, the outdoor enthusiast or the traditionalist. A sunken living room provides a lovely conversation area with a beautiful view of the great outdoors. An energy saving fireplace in the family room provides not only heat but a cozy atmosphere for those family gatherings. A well designed kitchen and dining room provides ease of entertaining both casual and formal. Grouped separate from the living areas are the bedrooms, the large master bedroom shows a dressing room and private full bath while 2 junior bedrooms share a centrally located bath. A hot tub highlights the large deck area accessible from either the family room or the master bedroom. A laundry links the garage with the kitchen.

**Living area—1,657 sq. ft.
(excluding garage and deck)**

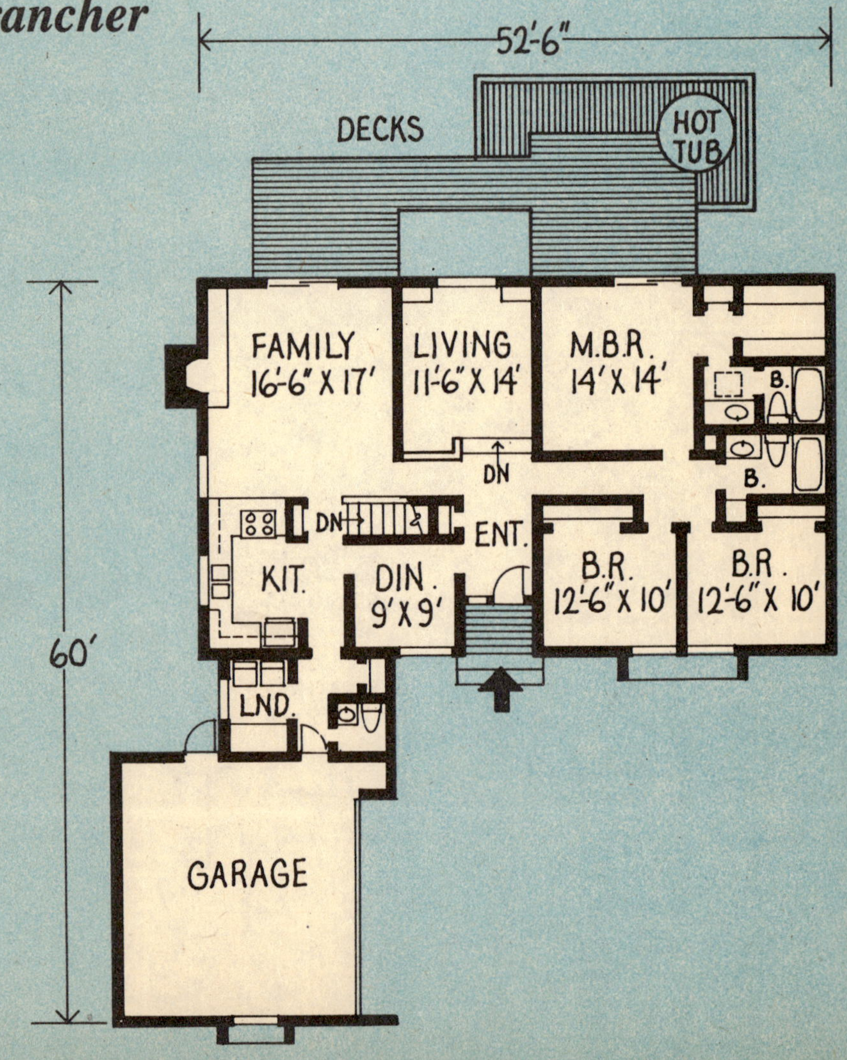

*For price and order information
see pages 108-109.*

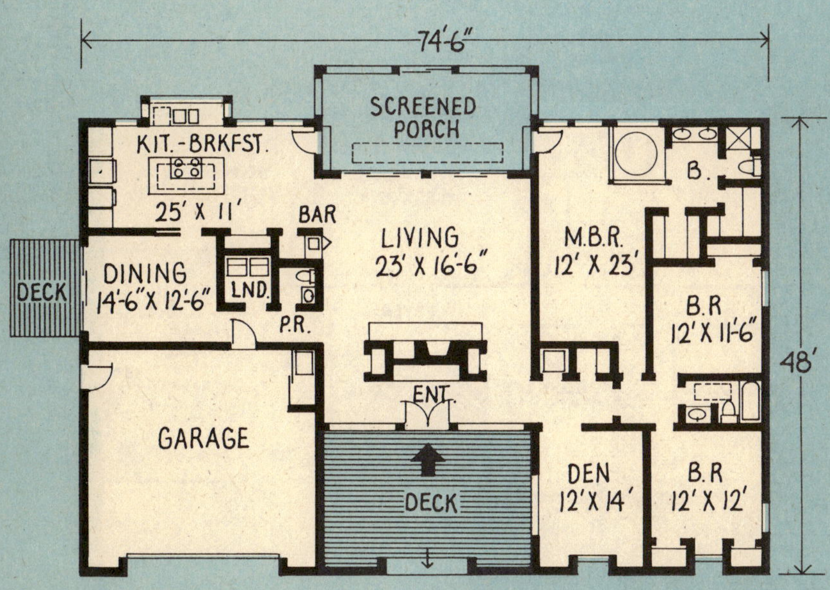

Low maintenance ranch

No. 26780—An exterior of fir siding with cedar battens provides beauty with age while little maintenance is required to preserve it's good looks. An inviting floor plan only adds to this well designed 4 bedroom home. The center of this home is a large living room with a stone fireplace. Sliding glass doors open from this area onto a screened porch. The bedrooms grouped at one end features a master bedroom complete with private bath, whirlpool and 2 large closets. The 3 smaller bedrooms share a centrally located bath. At the other end of the house the kitchen shows a center work island and breakfast area. The dining room opens to a deck for easy casual living. Other features include a laundry room, a lavatory and a bar. A large deck at the entry and a 2 car garage completes the picture.

Living area—2,550 sq. ft.

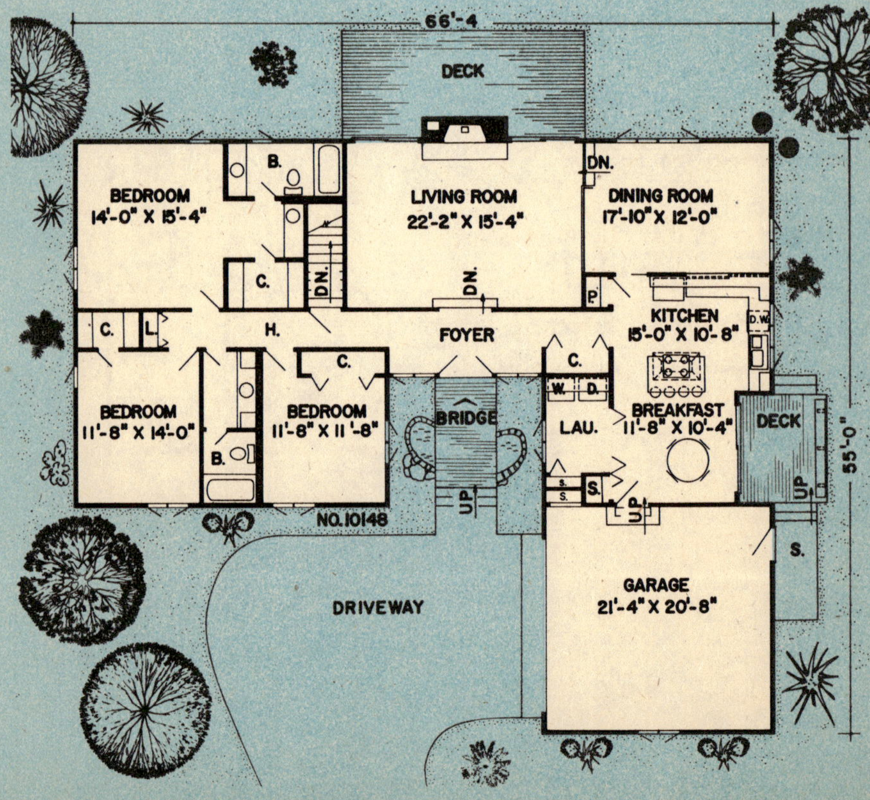

Redwood bridge fronts contemporary

No. 10148—Leading to double entrance doors and a lavish foyer, a redwood bridge expresses the unique, natural flavor of this three bedroom contemporary. Immediately visible from the foyer is the fireplace lighting the expansive sunken living room. Redwood deck beyond is accessible through two pairs of sliding glass doors.

First floor—2,050 sq. ft.
Basement—2,050 sq. ft.
Garage—440 sq. ft.

*For price and order information
see pages 108-109.*

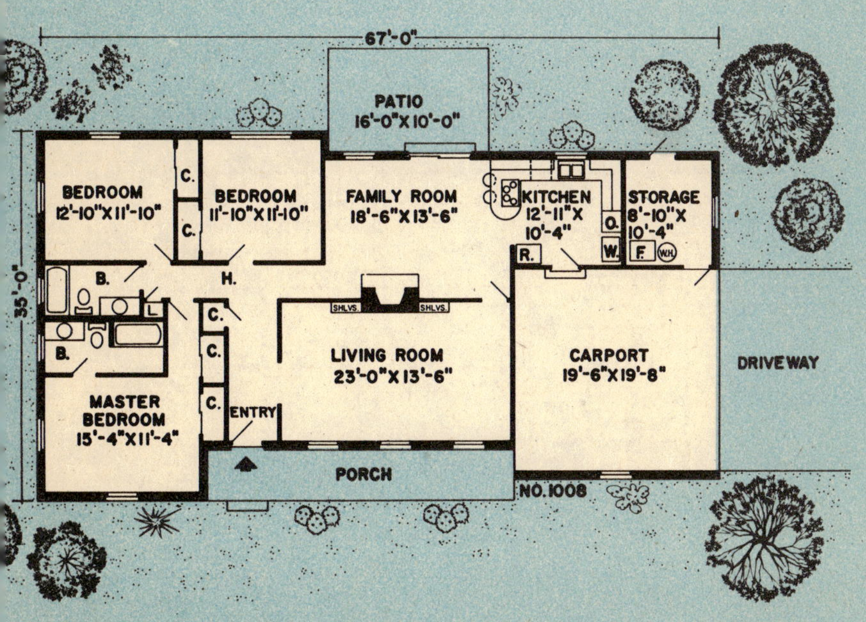

Plan boasts accommodating kitchen

No. 1008—Handy to patio, family room and living room, the kitchen in this home offers a snack bar and a garage entrance. Convenience marks the entire plan, which supplies a closeted entrance foyer to channel traffic. Warm and cozy, the family room is furnished with wood-burning fireplace. Three ample bedrooms share two full baths.

First floor—1,510 sq. ft.
Storage room—108 sq. ft.
Carport—417 sq. ft.

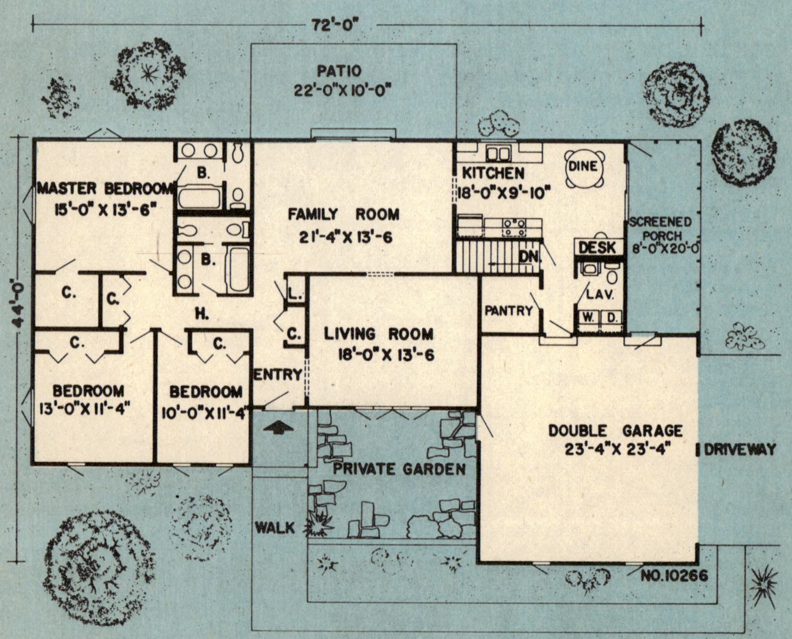

Screened porch extends dining area

**No. 10266—Outdoor living areas are
spotlighted in this one level plan, where
the living room overlooks a private
garden, the family room opens to a
patio, and the dining area is enlarged by
the adjacent screened porch. Also
exceptional is the kitchen complex, which
offers planning desk, dining area, and
pantry and borders a laundry/half bath.
Three bedrooms are outlined.**

**First floor—1,807 sq. ft.
Basement—1,807 sq. ft.
Porch—160 sq. ft.
Garage—539 sq. ft.**

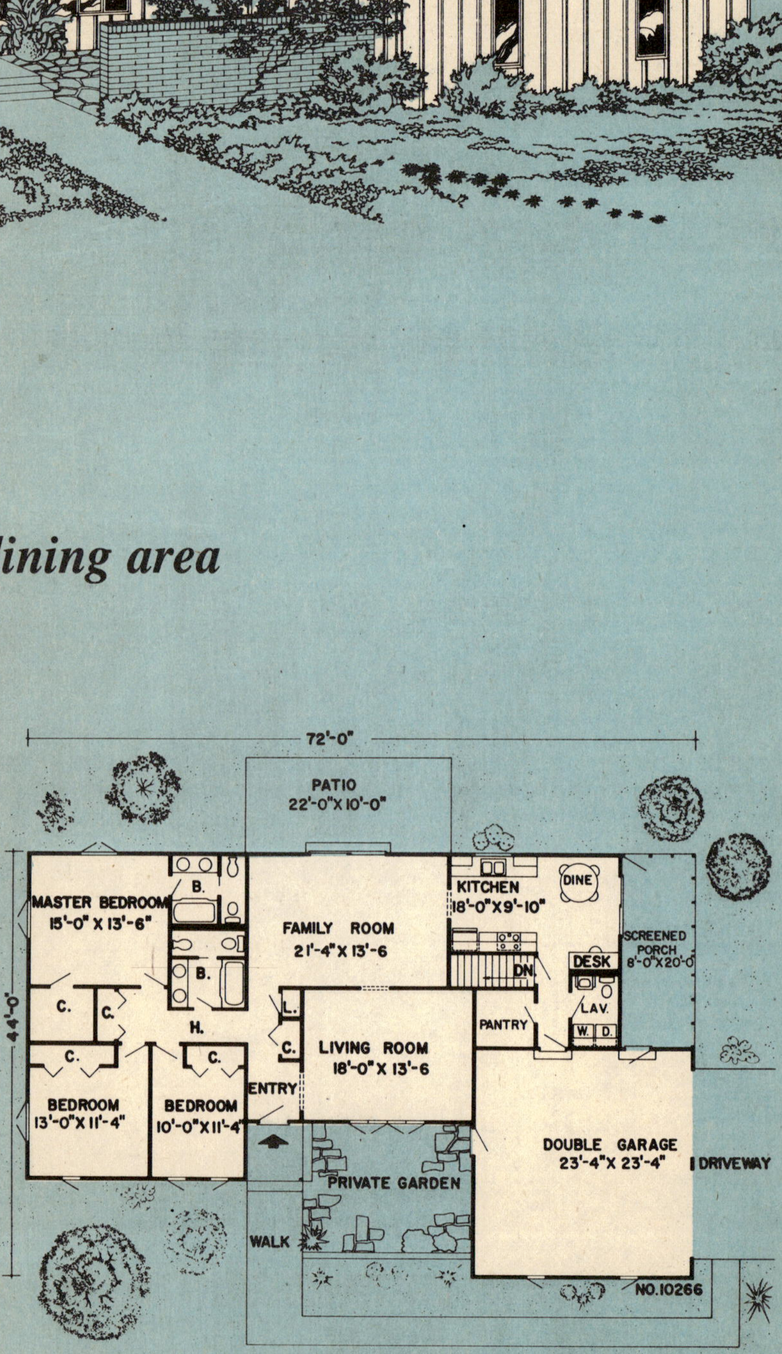

*For price and order information
see pages 108-109.*

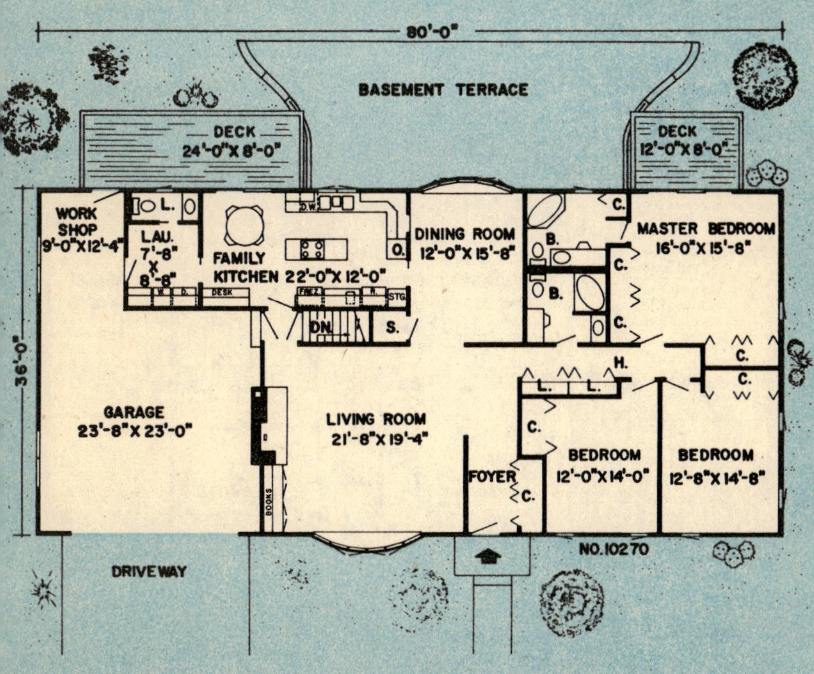

Master bedroom merits deck

No. 10270—Elegantly furnished with private bath and corner tub, the 16-ft. master bedroom in this appealing design opens to a private deck via sliding glass doors. The luxurious design also calls for bay windows in living room and dining room, desk and dining space in the kitchen, and handy workshop off the garage. A full basement, with attached terrace, is provided.

First floor—2,202 sq. ft.
Basement—2,016 sq. ft.
Garage/workshop—864 sq. ft.

Rustic ranch integrates outdoors

No. 10142—Appendaging a 31-foot redwood deck at rear and a long front porch, this ranch plan offers a woodsy appeal and plenty of involvement with the outdoors. Inside, the floor plan caters to the relaxed lifestyle of the seventies. Flanking the large foyer is the spacious sunken living room, warmed by a wood-burning fireplace.

First floor—1,705 sq. ft.
Basement—1,705 sq. ft.
Garage—576 sq. ft.

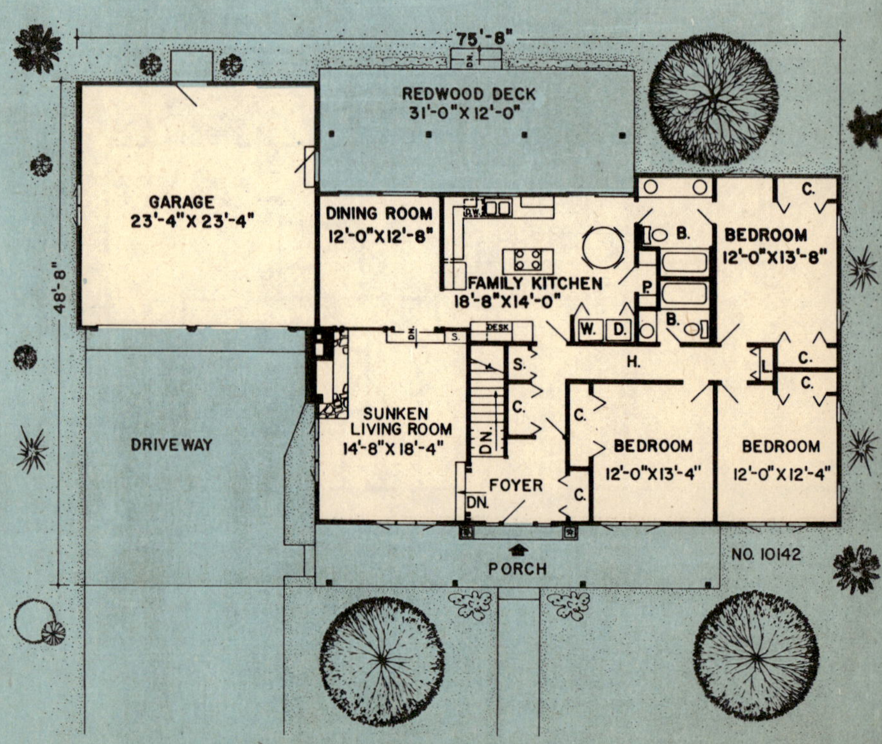

For price and order information see pages 108-109.

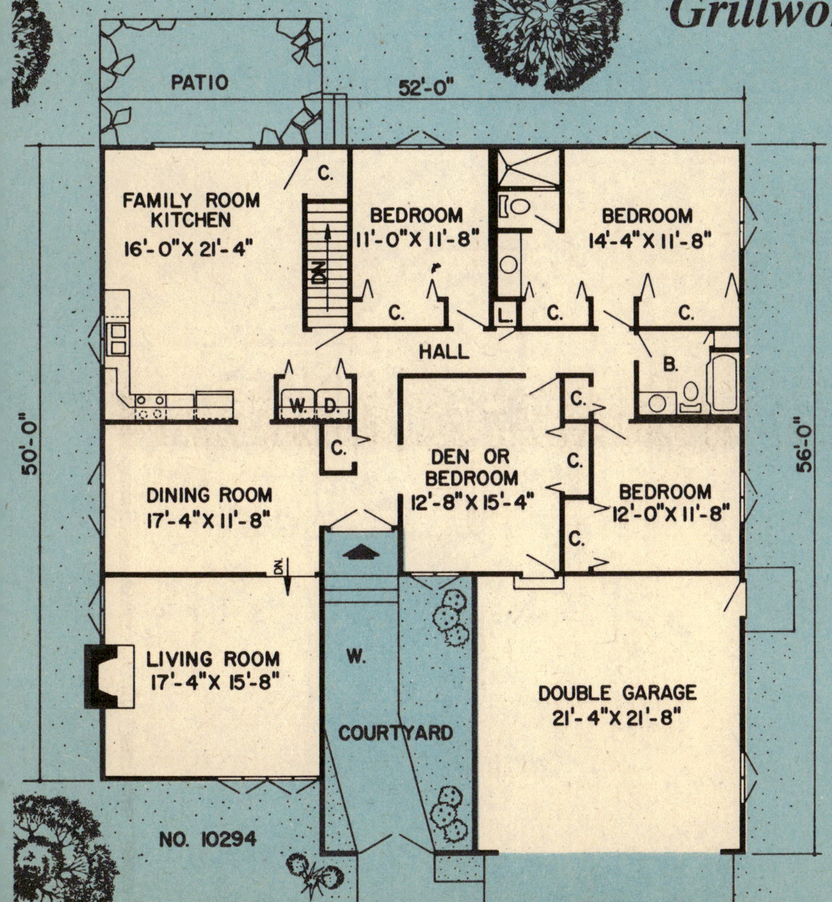

Grillwork, courtyard, add interest

No. 10294—Designed to maximize space and appeal, this four bedroom one level plan begins with a private courtyard introduced by arched grillwork. The sunken living room offers a fireplace and a dead-end arrangement, and dining space is found in the formal dining room as well as the 21-ft. family kitchen. With garage and hall access, the fourth bedroom can serve as guest room, library, or hobby area.

First floor—2,034 sq. ft.
Basement—2,034 sq. ft.
Garage—484 sq. ft.

Fireplace center of circular living area

No. 10274—A dramatically positioned fireplace forms the focus of the main living area in this single level contemporary. Kitchen, dining, and living rooms form a circle that allows work areas to flow into living areas and outdoors, via sliding glass doors, to the wood deck. Three large bedrooms and two full baths are grouped in a wing away from the living areas. Tucked off the kitchen is a laundry room with closet.

Living area—1,783 sq. ft.
Garage—576 sq. ft.

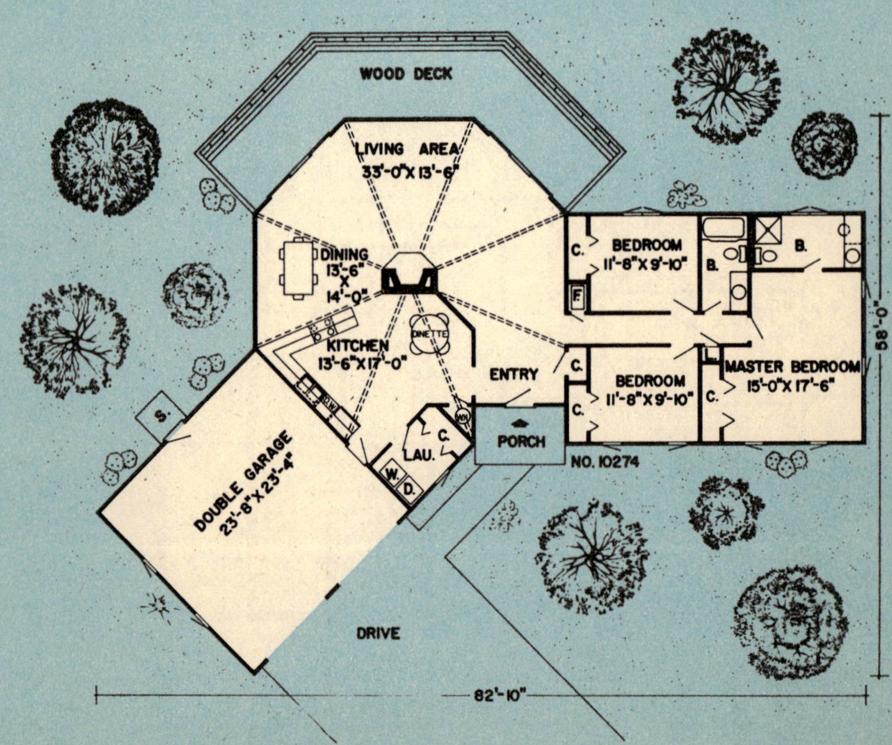

For price and order information see pages 108-109.

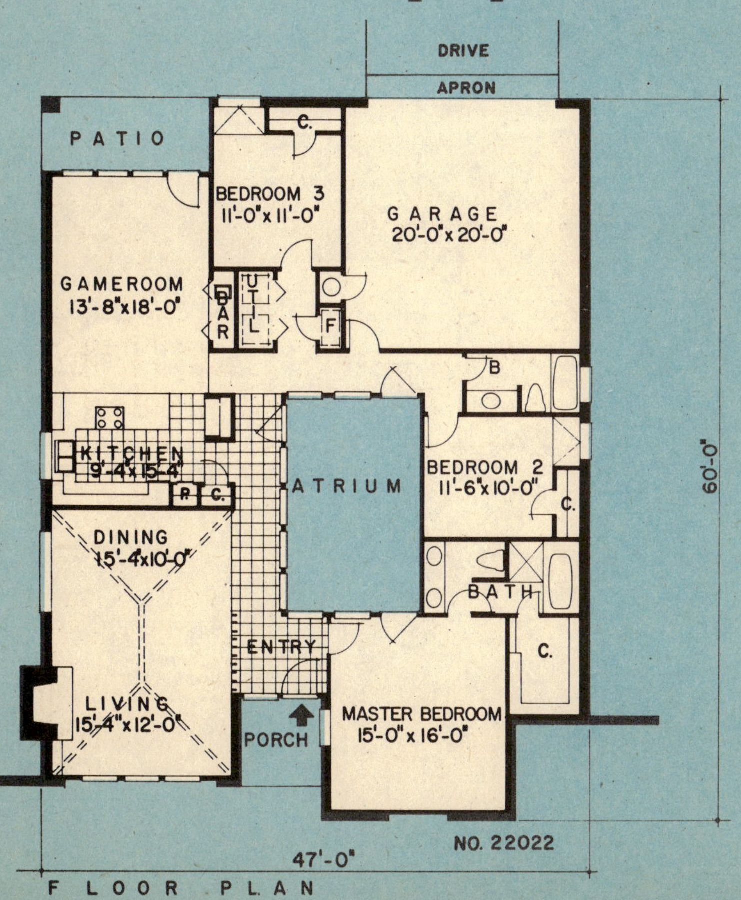

Unique plan encircles atrium

No. 22022—Functional in design, this three bedroom home is enhanced by a central atrium, a restful retreat accessible from hallways and master bedroom. The U-shaped kitchen calls for a built-in pantry, and the neighboring game room includes a wet bar. For family and friends, the spacious living and dining room merits a wood-burning fireplace.

House proper—1,913 sq. ft.
Garage—441 sq. ft.
Atrium—204 sq. ft.

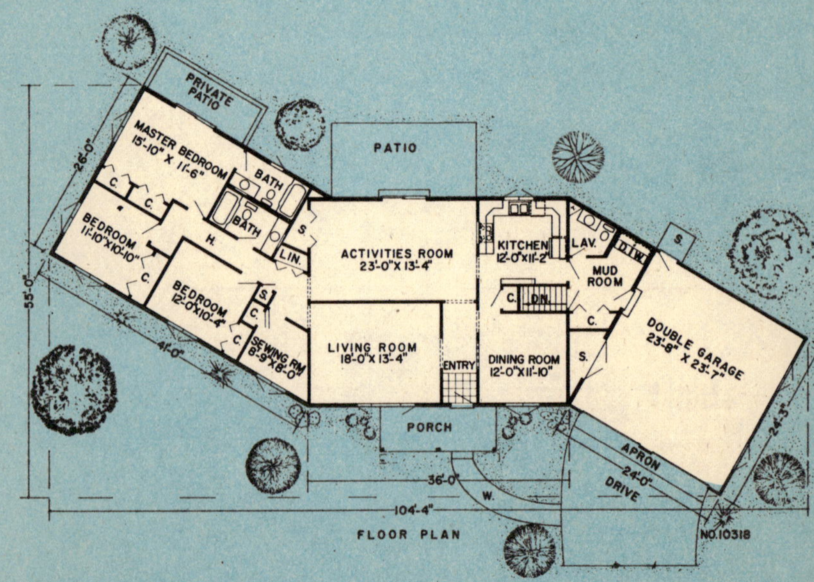

Activity wings define ranch plan

No. 10318—Three wings effectively separate sleeping areas, activity areas, and garage in this three bedroom ranch style. In the sleeping wing, large bedrooms are grouped with baths and sewing room, while the main wing offers formal living and dining rooms backed by kitchen and activities room. Two patios are featured.

First floor—2,082 sq. ft.
Garage—583 sq. ft.
Basement—1,008 sq. ft.

For price and order information see pages 108-109.

A-Frame inspires striking design

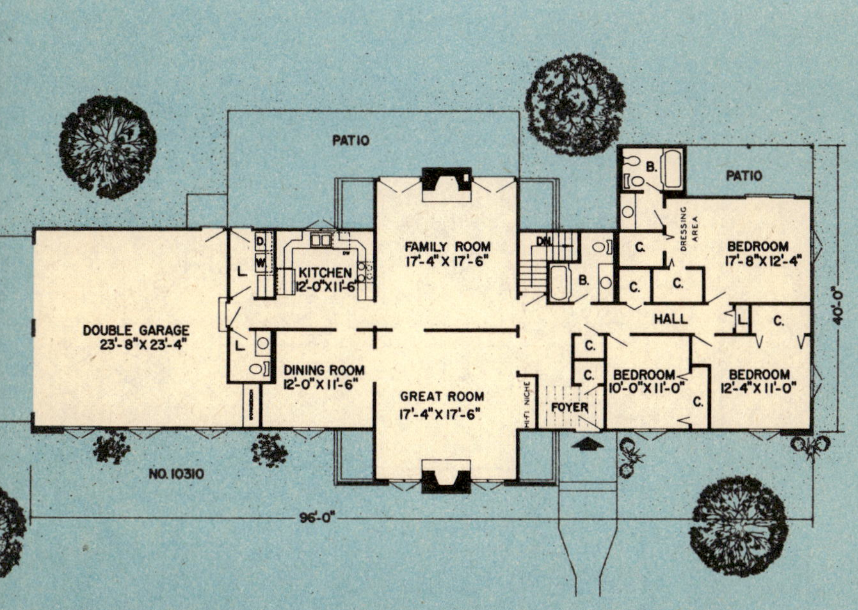

No. 10310—With a facade that suggests a winged A-frame home, this one level plan achieves uniqueness and livability. Notable is the commanding great room with wood-burning fireplace, backed by a firelit family room of identical size. The master bedroom boasts a dressing area, walk-in closets, and private bath, as well as sliding glass doors to its own patio.

First floor—2,090 sq. ft.
Basement—2,090 sq. ft.
Garage—576 sq. ft.

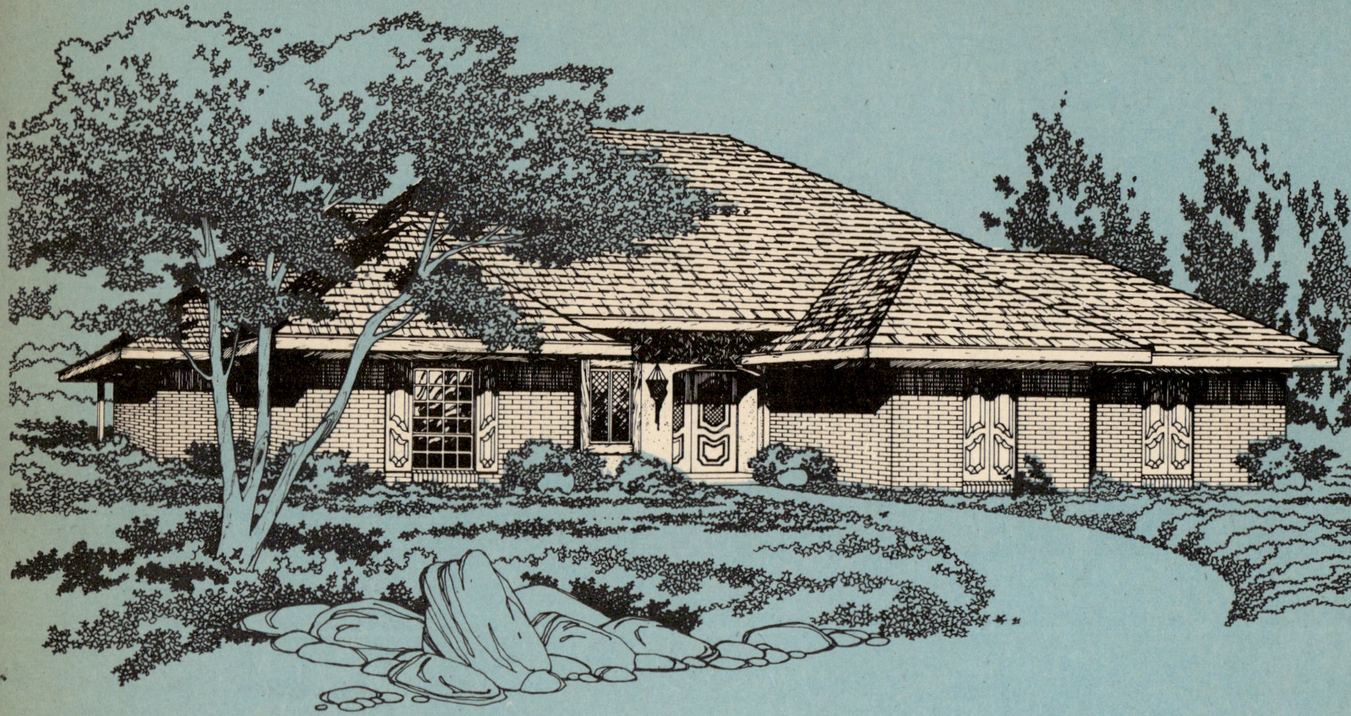

One level home offers utility room, pantry

No. 22000—Inside its gracious hip roof exterior, this single level plan shows well-defined living areas and such extras as a utility room off the garage and a pantry off the kitchen. Wherever possible, outdoor living areas are merged with those indoors, as in the game room, family room, and dining room, each of which overlook or open to patios. A closet, wet bar, and powder room furnish the game room area.

House proper—2,094 sq. ft.
Garage—498 sq. ft.

FLOOR PLAN NO. 22000

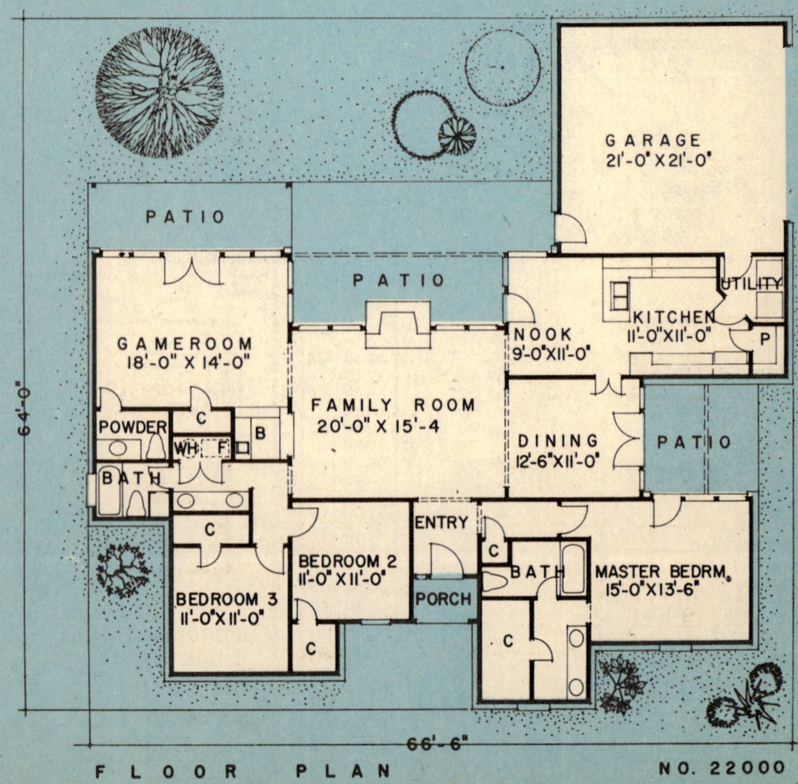

For price and order information see pages 108-109.

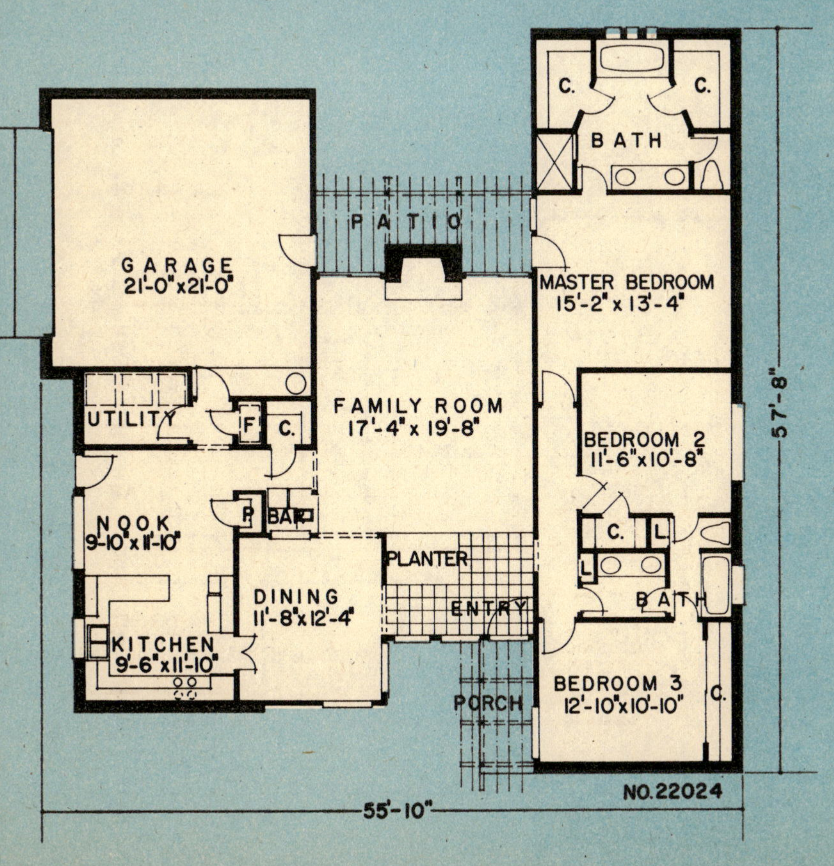

Skylit family room focus of plan

No. 22024—Directly accessible from the airy foyer, the 19-ft family room offers a center for family activity as well as a skylight, fireplace, and sliding glass doors to the adjoining patio. This single level plan groups three bedrooms and two full baths and shows a utility room and dining nook. A wet bar edges the dining and family rooms.

House proper—1,974 sq. ft.
Garage—505 sq. ft.

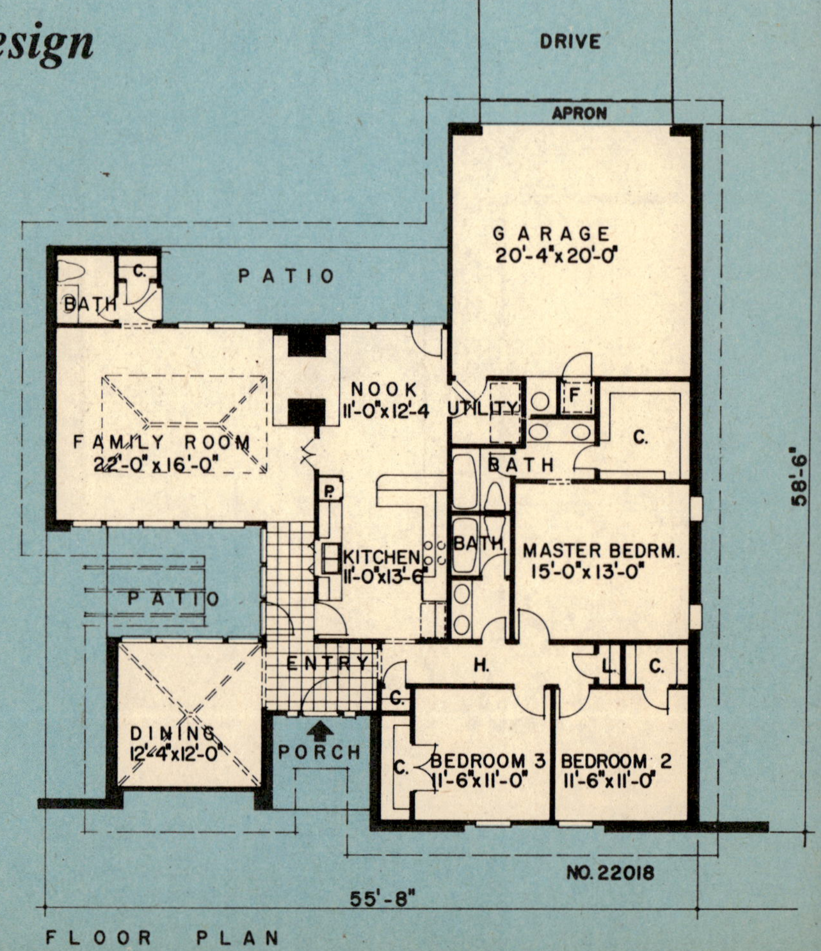

Effective zoning marks design

No. 22018—Bedrooms, family living areas, and formal dining room are carefully defined and separated in this well-zoned plan. Three bedrooms cluster around two full baths at right, while the left wing spotlights a family room, hallway, and dining room that encircle and overlook the patio. Closets are plentiful throughout.

House proper—1,879 sq. ft.
Garage—471 sq. ft.

*For price and order information
see pages 108-109.*

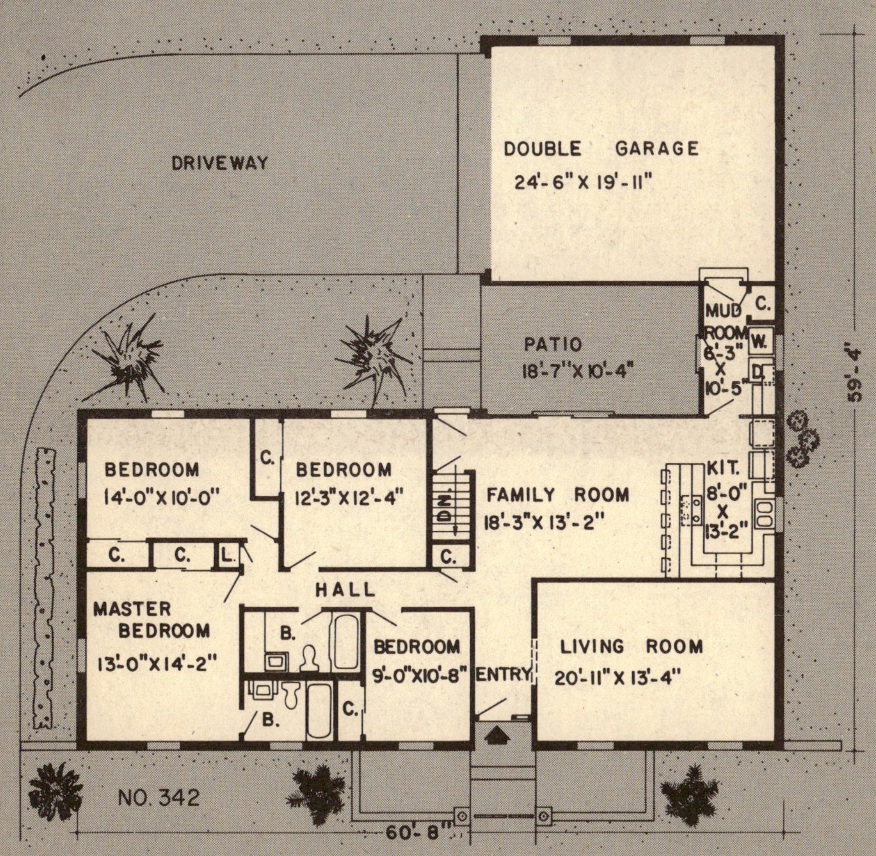

Private patio highlight home

No. 342—Planned for convenience and livability, this four bedroom design offers an 18-foot patio, enclosed on three sides for maximum privacy. Adjoining family room and kitchen share a snack bar, and the combination mud room/laundry can be reached via garage, patio, or kitchen.

First floor—1,869 sq. ft., Basement—1,703 sq. ft. Garage—546 sq. ft.

For price and order information see pages 108-109.

Design symmetry enhances appearance

No. 366—Off at one end of the house, noise made in the family room or garage will not interrupt concentration or sleep in the bedrooms. While children's activities may be confined to the family room, both the living room and dining room are designed for daily use and family lounging. A fireplace in the living room and barbecue grill adjacent to it on the patio add the spark of warmth desired by so many of today's families.

First floor—1,925 sq. ft., Basement—1,925 sq. ft. Garage—473 sq. ft.

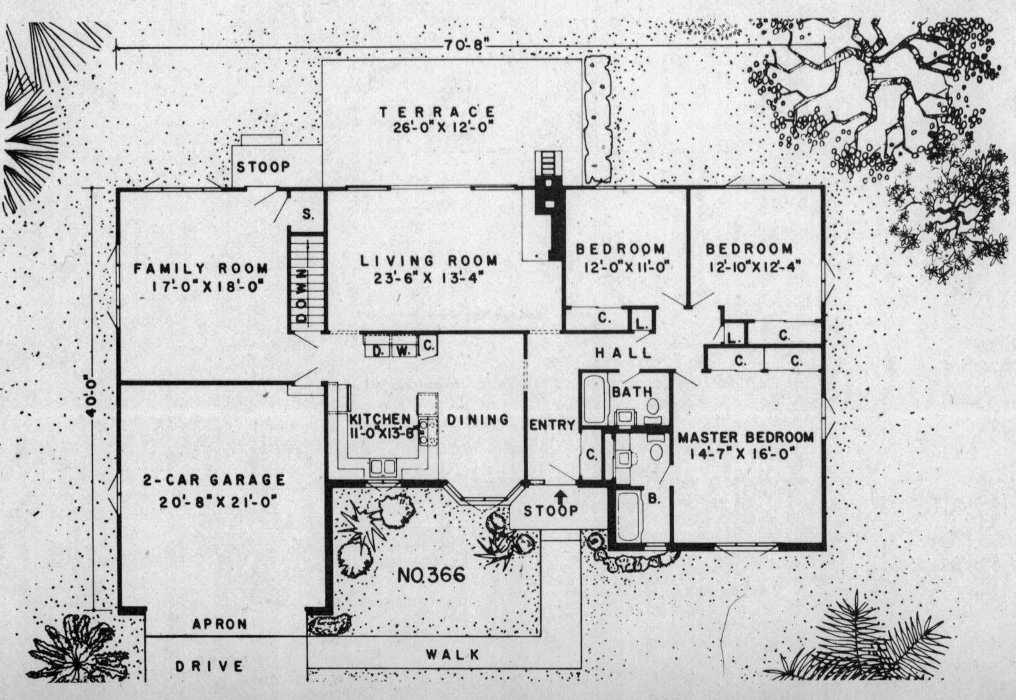

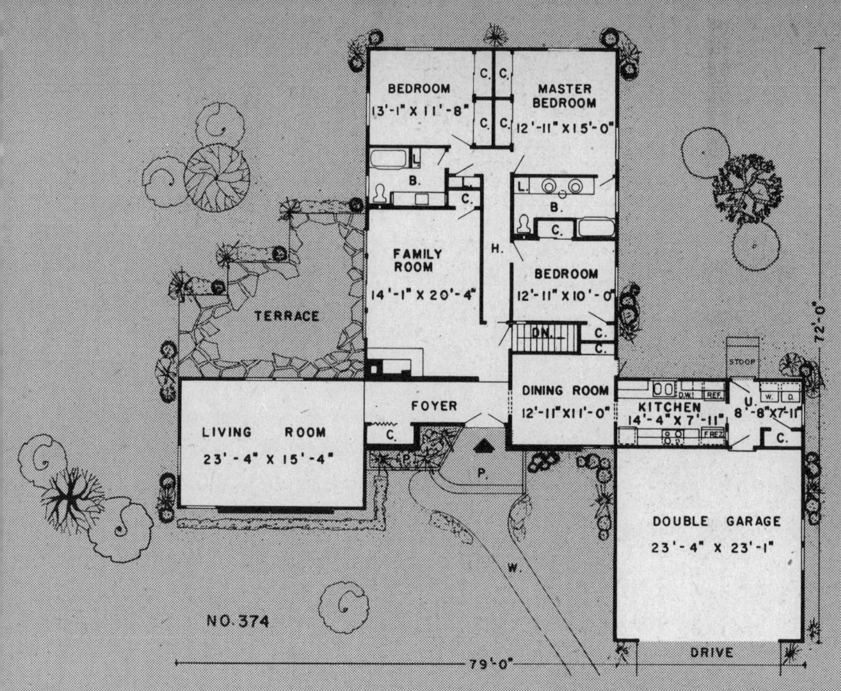

Triple-winged plan expands for living

No. 374—Three distinct areas stretch to capture space and zone areas for family living in this well-arranged contemporary. Enjoying a corner of its own and a view of the terrace, the living room spans over 23 feet. To the right of the closeted foyer, a formal dining room connects with corridor kitchen and utility room.

First floor—2,107 sq. ft., Basement—1,290 sq. ft. Garage—562 sq. ft.

Story book house for narrow lot

No. 356—Old English brick and vertical redwood siding are pleasingly combined on the exterior of this compact house. The foyer is located to provide easy access to all areas of the house. A centrally located kitchen makes it convenient to all members of the family while allowing a minimum of steps for the housewife. Attaching the garage to the rear of the house keeps the overall width under 50 feet making it well suited for narrow lots.

First floor—1,610 sq. ft., Garage—593 sq. ft.

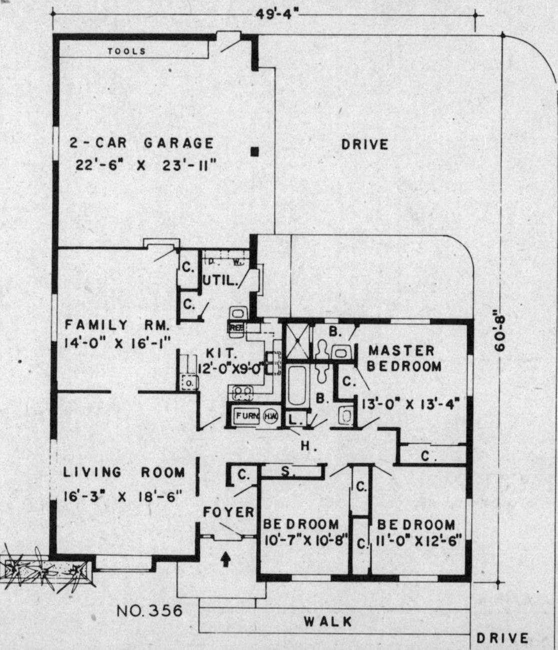

*For price and order information
see pages 108-109.*

Country kitchen boasts barbecue grill

No. 9830—Quaint and colorful, the country kitchen of this hipped-roof home is favored with a built-in barbecue and a breakfast bar that separates it from the family room. Carefully detailed, the plan assigns the family room sliding glass doors to take advantage of the triangular terrace and the well-windowed living room a pleasurable wood-burning fireplace. Bedrooms are placed to the rear of the design and include a nicely-proportioned master bedroom with bath, two more substantial bedrooms, and a slightly smaller den.

First floor—1,782 sq. ft., Basement—1,782 sq. ft. Garage—576 sq. ft.

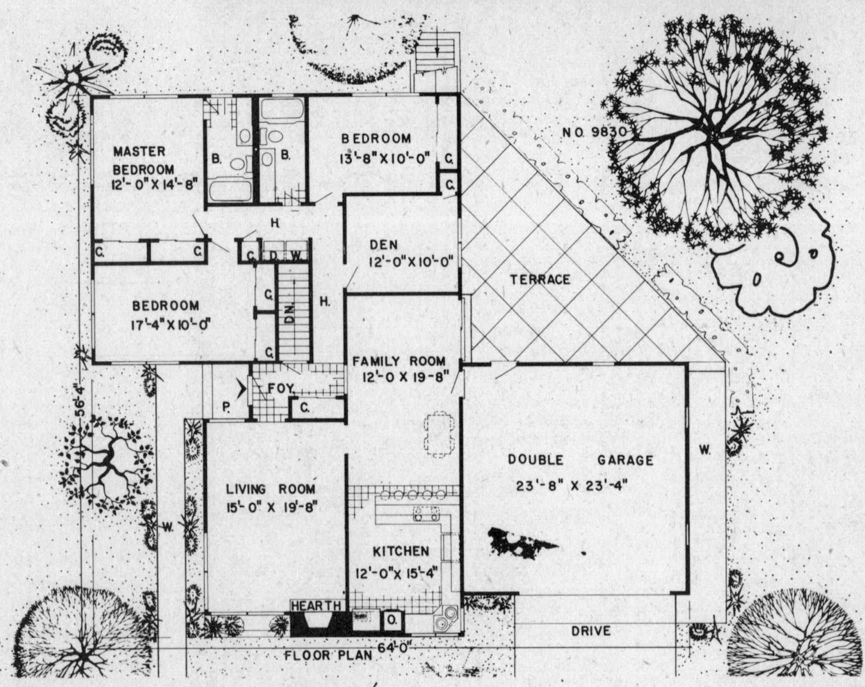

Home recalls Southern plantation

No. 9850—Magnificent white columns, shutters, and small paned windows combine to create images of the pre-Civil War South in this generously-proportioned design. Inside, the opulent master bedroom suite, with plentiful closet space, a full bath and study, suggests modern luxury. Fireplaces enhance the formal living room and sizable family room, which skirts the lovely screened porch. The formal dining room boasts built-in china closets.

First floor—2,466 sq. ft., Basement—1,447 sq. ft.
Garage—664 sq. ft.

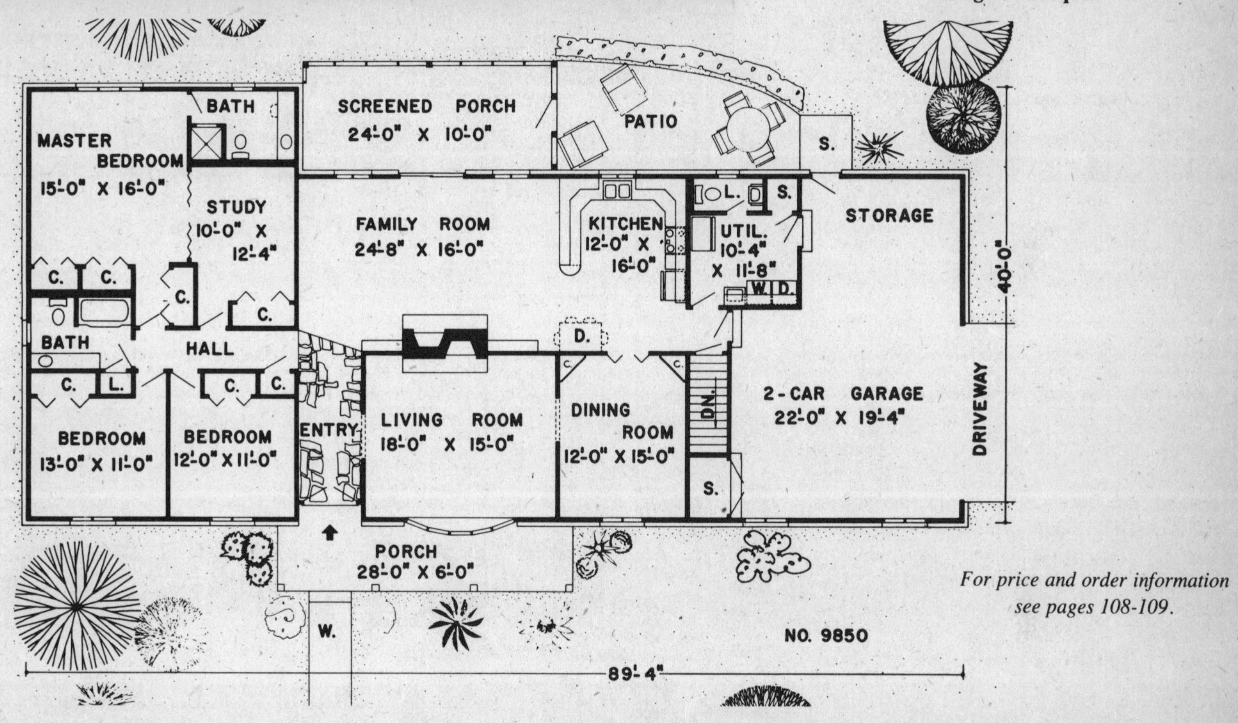

For price and order information
see pages 108-109.

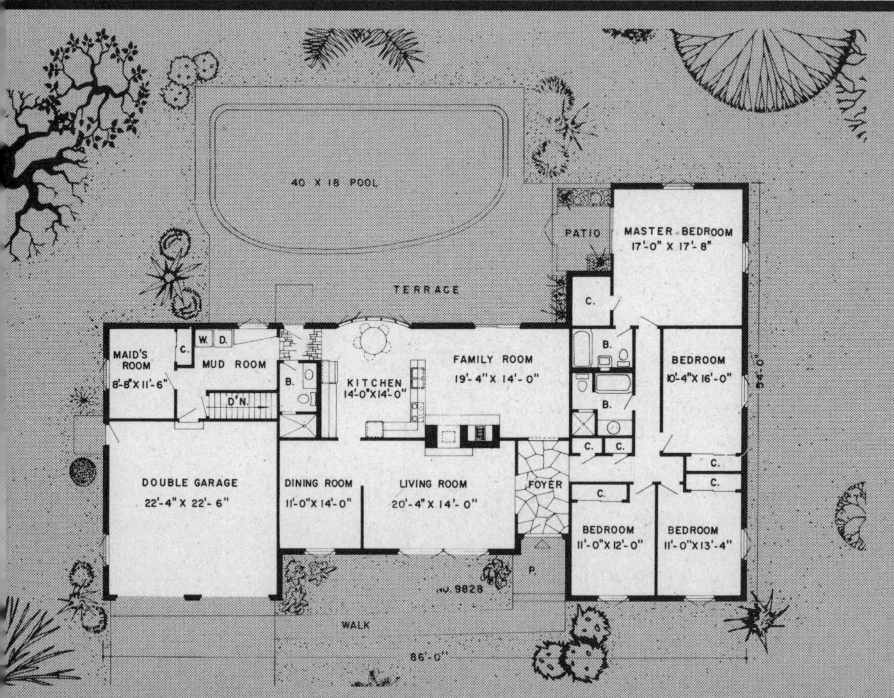

Come on in, the living's fine

No. 9828—This house leaves little to be desired in a comfortable family home. A few of the most outstanding features are a private patio off of the master bedroom; a two-way fireplace between the living room and family room; a built-in charcoal grill in the family room; a bath with shower next to the kitchen and handy to the pool area and a very nice breakfast nook with a view of the pool area. The mud room is large enough to serve as a sewing and ironing room as well as the laundry. The low maintenance exterior of beautiful natural stone blends well with the shake shingle roof.

First floor—2,679 sq. ft., Basement—2,679 sq. ft.
Garage—541 sq. ft.

Sleeping wing separate, restful

No. 386—Maximum quiet and privacy result from the secluded layout of the bedroom wing in this charming ranch style. Four bedrooms are nestled around two full baths, with linen closets and laundry niche adding convenience. Informal family room offers access to basement, terrace, and kitchen. Guests can be accommodated in the 21-foot living room, complete with corner fireplace. For garden equipment, bicycles, and tools, substantial storage space is provided in the double garage.

First floor—1,649 sq. ft., Basement—945 sq. ft. Garage—520 sq. ft.

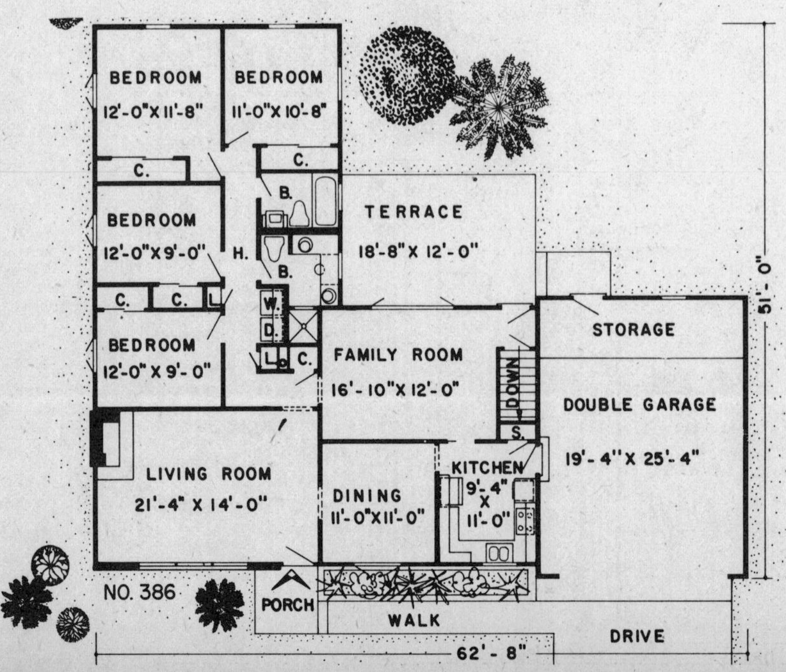

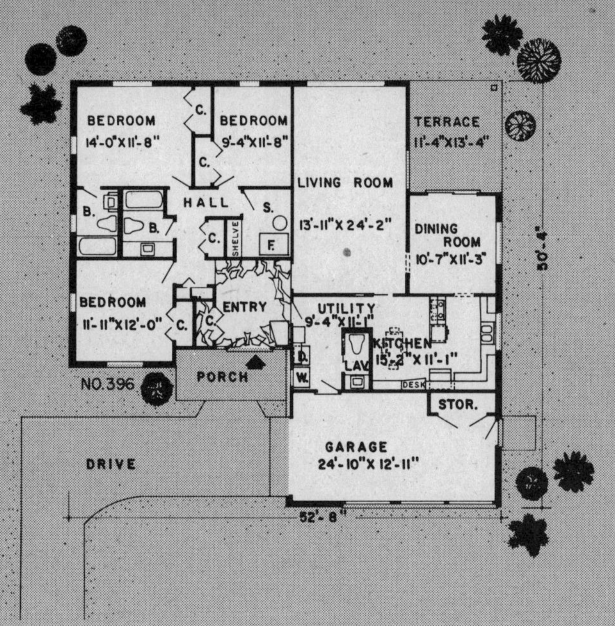

Double doors create inviting facade

No. 396—Welcoming guests into the flagstone foyer, the double doors of this attractive home create an inviting atmosphere that is carried through on the interior. An efficient floor plan places utility room and eat-in kitchen at right, with living and dining rooms adjoining. Open to the foyer, the 24-foot living room joins the formal dining room to provide party space, and the dining room annexes the terrace via sliding glass doors. The bedroom wing benefits from two full baths and ample closet space, including a large utility closet with built-in shelves.

First floor—1,686 sq. ft., Garage—360 sq. ft.

Kitchen proposes relaxed informality

No. 390—Featuring four entrances and an eye-catching angular snack bar, the expansive kitchen-dining area suggests a gathering spot within this three bedroom home. For more formal moments, the design also offers a 24-foot living room with mood set by crackling corner fireplace and sliding glass doors to the terrace. The master bedroom promises comfort and rates double closets and large, private bath with two sinks and towel closet. Sizable utility room bordering the kitchen will solve most storage problems.

First floor—1,637 sq. ft., Garage—420 sq. ft.

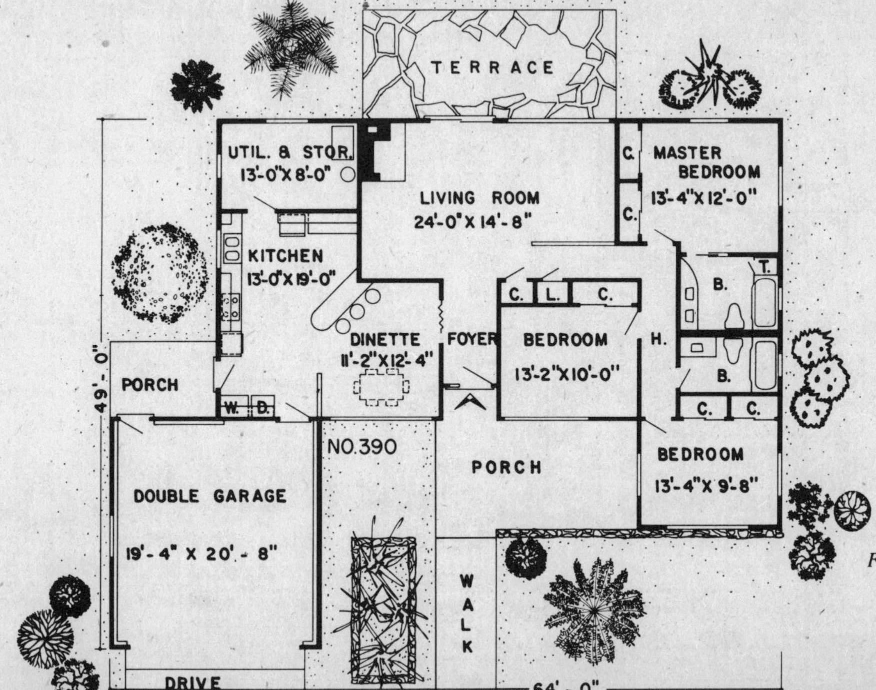

*For price and order information
see pages 108-109.*

Wealth of windows lights design

No. 384—Generous expanses of glass crowned with cathedral ceilings emphasize light and space in this airy contemporary plan. Entry allows immediate access to the dining room and firelit living room which overlooks the terrace. Kitchen and family room form an open area with snack bar and terrace entry, and adjoining utility room boasts laundry space and half bath. Two full baths, each with towel closet, and four sizable bedrooms make up the sleeping wing. Master bedroom merits private bath, double closets.

**First floor—2,118 sq. ft., Basement—1,817 sq. ft.
Garage—476 sq. ft.**

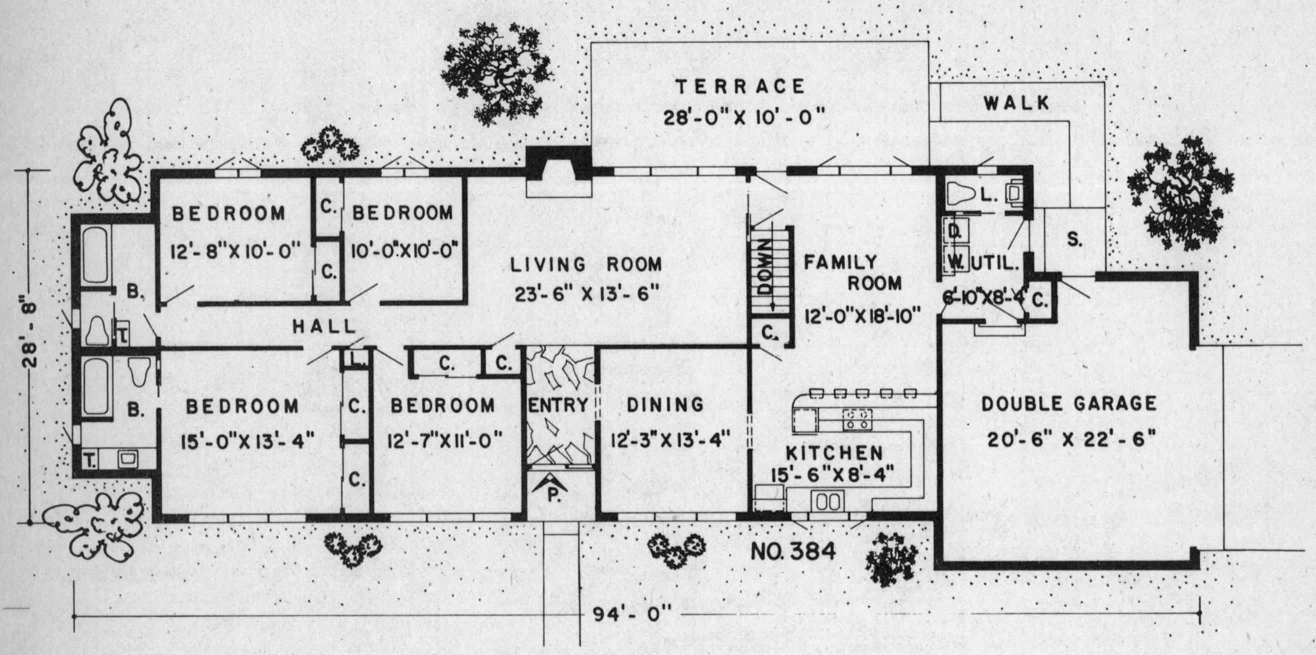

Shade, access benefit ranch style

No. 392—Shaded by a wide roof overhang, this unique ranch style offers a floor plan that virtually eliminates cross-traffic, with ready access to bedrooms, living room, and family room. The design generously places closets throughout the sleeping area and shows "his" or "hers" closets and private bath for the master bedroom. Entry to the dead-end living room is preceded by a coat closet, and a adjoining family room. Etching a laundry corner and plentiful counter space, the long kitchen opens to an oversized double garage and storage area.

First floor—1,581 sq. ft., Garage—582 sq. ft.

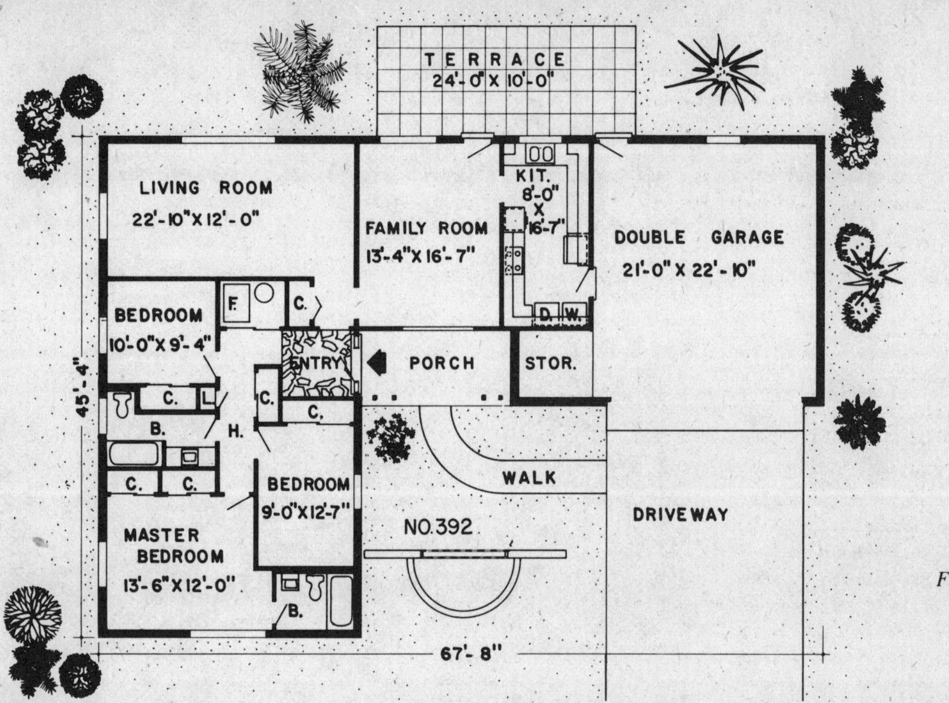

For price and order information see pages 108-109.

Traditional boasts foyer, closet space

No. 388—Plentiful closet space favors bedrooms and foyer in this traditionally detailed plan. Arranged for maximum use of space, the floor plan shows a sizable living room off the foyer and a family room-kitchen combination which overlooks and opens to the terrace. A snack bar serves the area, and bordering 12-foot laundry room is designed to save steps. Three accessible bedrooms are well-closeted and two full baths are placed back to back for construction economy. Additional storage and workshop space abounds in basement and double garage.

First floor—1,602 sq. ft., Basement—1,290 sq. ft. Garage—574 sq. ft.

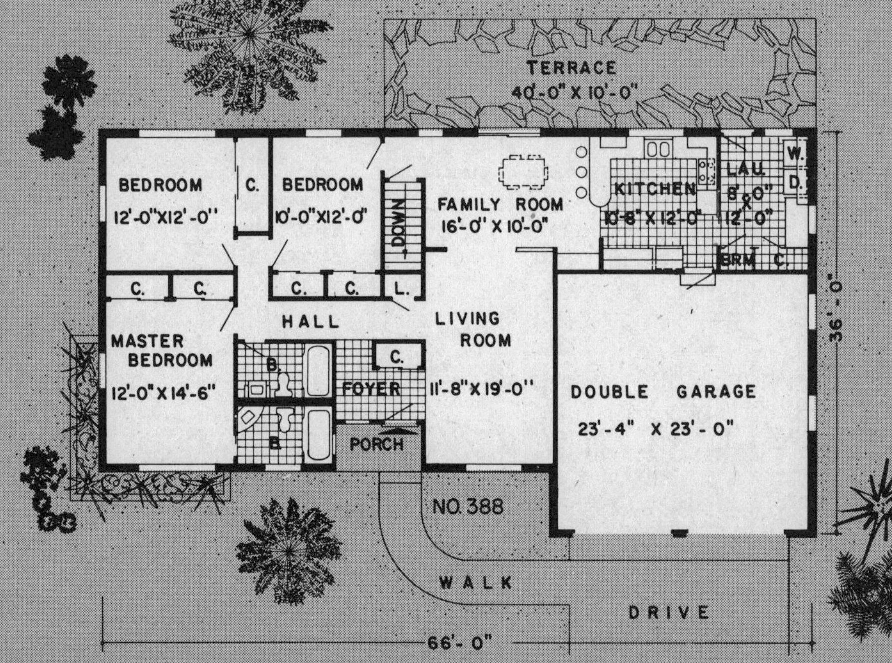

Cathedral ceilings striking decor

No. 9058—Visitors to this contemporary home are immediately struck by the awe-inspiring cathedral ceilings of the foyer, living and dining rooms and kitchen. The living room, recessed a foot below the foyer and dining room, has access to the terrace through sliding glass doors. Entrance to the terrace also is provided through the family and dining room. Three large bedrooms and two and one-half baths are planned to accommodate the large family, and the convenient mudroom provides laundry and closet space, plus a half-bath.

First floor—1,970 sq. ft., Garage—526 sq. ft.

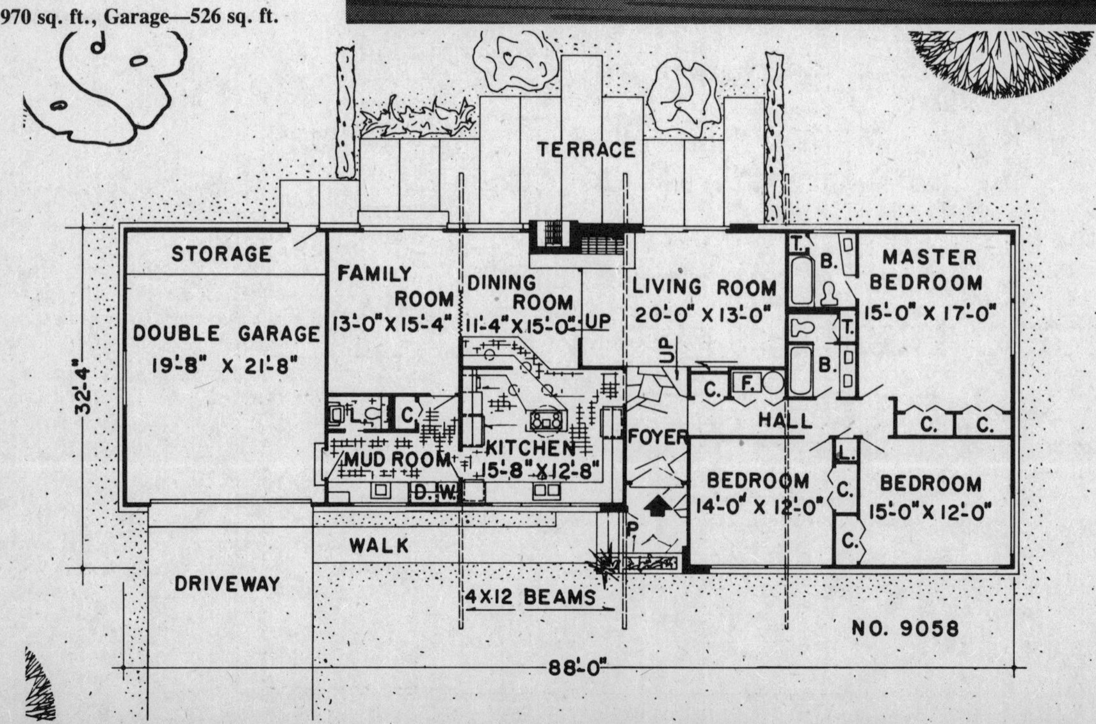

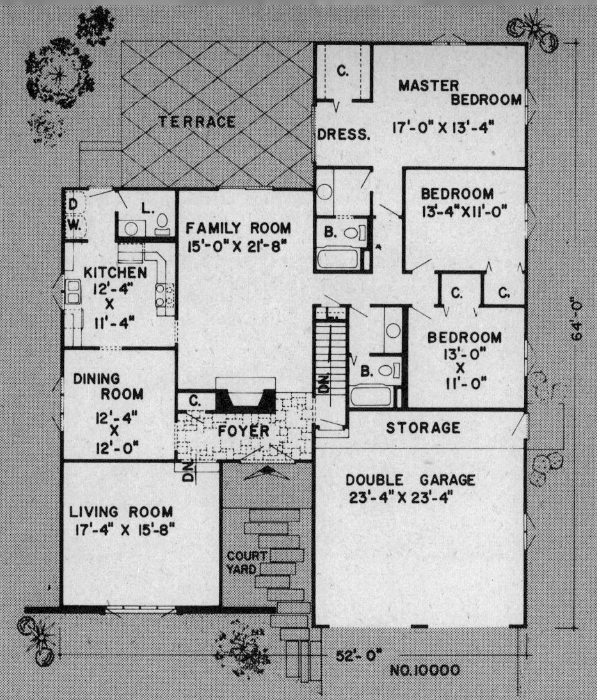

Courtyard provides impressive entry

No. 10000—Outstanding Mediterranean style homes, like this one, are in demand. The courtyard has massive beams overhead which make a very impressive entrance. The stone arch over the living room window, arched garage doors, exposed gable and beams and textured battened siding add further to the beauty of this home. The floor plan is equally outstanding. The formal living room is sunken, has cathedral ceilings and several large exposed beams. The paneled family room has an extra large wood-burning fireplace and sliding glass doors which open onto a large terrace. There are three bedrooms and 2½ baths. The master bedroom is quite elegant.

First floor—2,103 sq. ft., Basement—1,807 sq. ft. Garage—569 sq. ft.

Cathedral ceilings climax contemporary

No. 9994—Airy and exciting, the sunken living room in this clean contemporary celebrates via high-reaching beamed ceilings, a wood-burning fireplace, and double sliding glass doors to the terrace. To its right, the formal dining room slides open to a screened porch, which also joins the breakfast nook. Bordering the substantial family room is a full bath with shower. Four generous bedrooms produce a sharply-planned sleeping area with minimum waste space. Bountiful closet space, segmented bath, and dressing area benefit the master bedroom.

First floor—2,722 sq. ft., Basement—1,210 sq. ft. Garage—528 sq. ft.

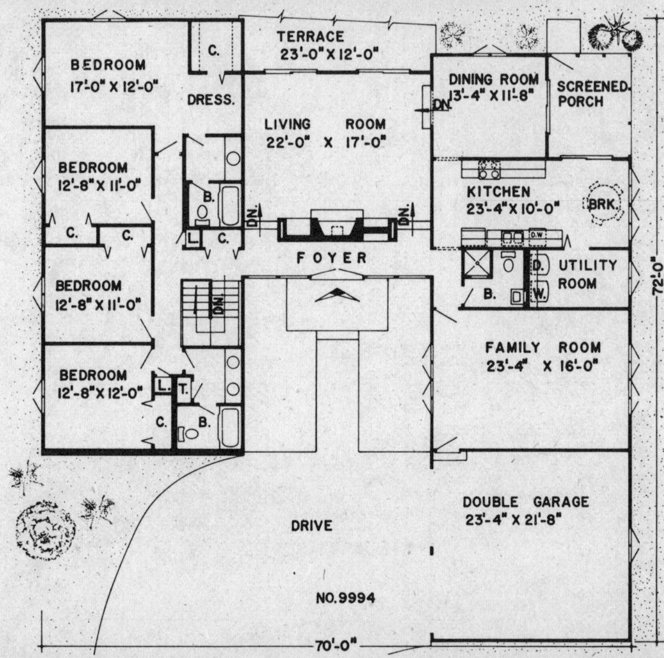

*For price and order information
see pages 108-109.*

Contemporary plan angular, unique

No. 310—Multi-angular and striking, the family room-kitchen area of this contemporary plan serves as cornerstone and focus. Corner fireplace spreads a glowing warmth, while an island snack bar expresses contemporary convenience. Dead-end living room and formal dining room wing the flagstone foyer, which allows ready access to sleeping quarters. Three large, well-closeted bedrooms are provided with two full baths.

First floor—2,132 sq. ft., Garage—606 sq. ft.

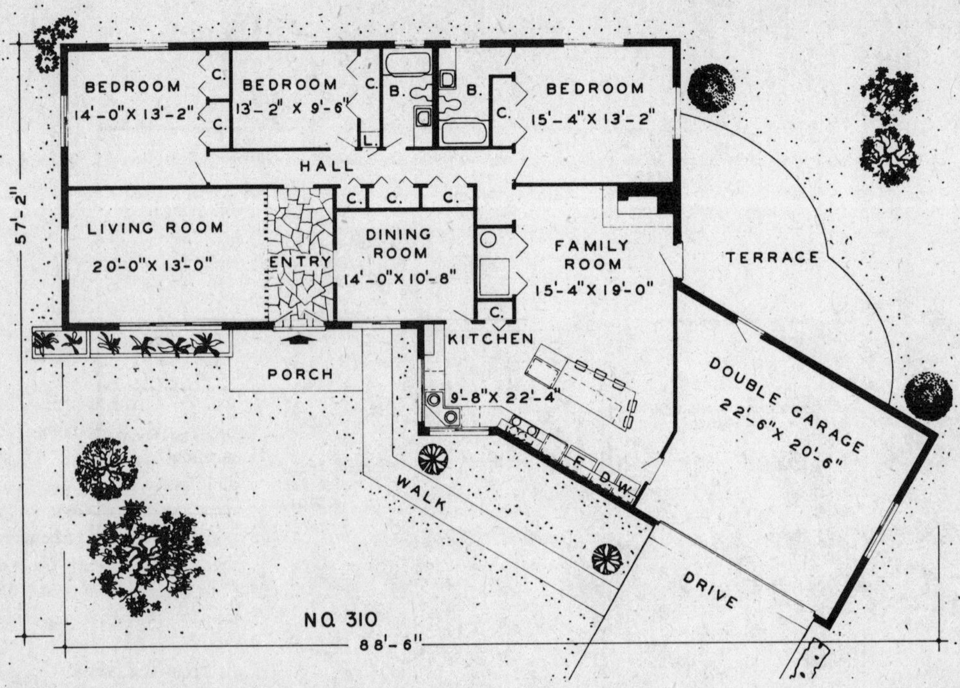

Kitchen complex noted for efficiency

No. 160—Sporting both dining and laundry centers, the kitchen complex in this appealing ranch style serves as a functional unit, handy to living room and family room. Sleeping areas are clearly defined for privacy, especially the separate master bedroom with bath. Tiled entry and hallways channel traffic and allow the living room seclusion and formality. Set behind the garage, the family room provides a large recreation area for active use and opens to the outdoors via sliding glass doors.

First floor—1,782 sq. ft.
Basement—1,782 sq. ft.
Garage—406 sq. ft.

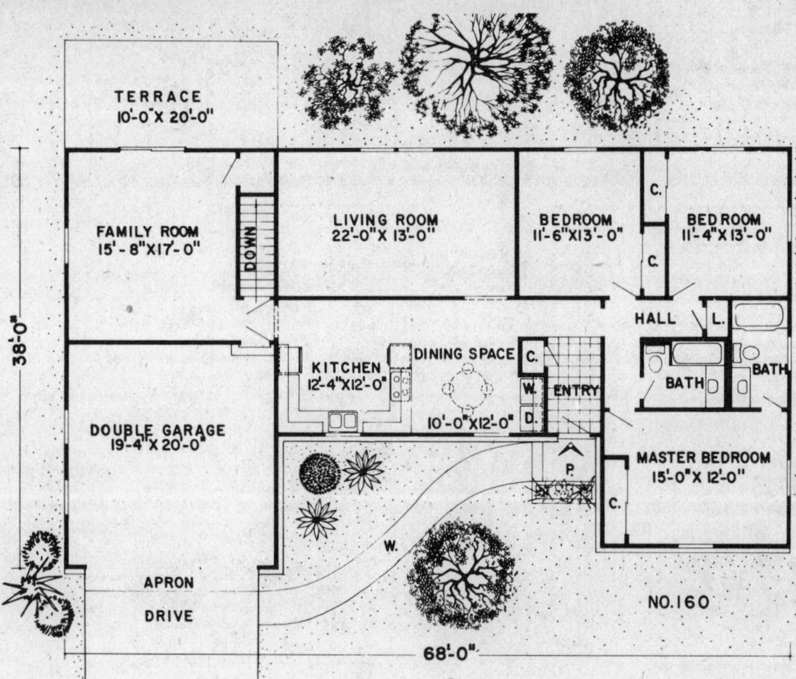

*For price and order information
see pages 108-109.*

An air of elegance

No. 242—A sunken living room is featured in this beautiful hipped roof, ranch style house. The entrance foyer is very attractive with its tiled floor and louvered partition next to the breakfast room. The foyer connects to a central hall which efficiently channels traffic to each room. The master bedroom has its own private bath, wood-burning fireplace and walk-in closet. Open planning is the rule in the family room, kitchen area. A long snack bar serves as a divider. The half bath and laundry room are conveniently located near the kitchen area. A barbecue grill is included in the family room, utilizing the living room fireplace chimney. Both living room and family room open onto the covered patio through sliding glass doors.

First floor—2,227 sq. ft., Garage—643 sq. ft.

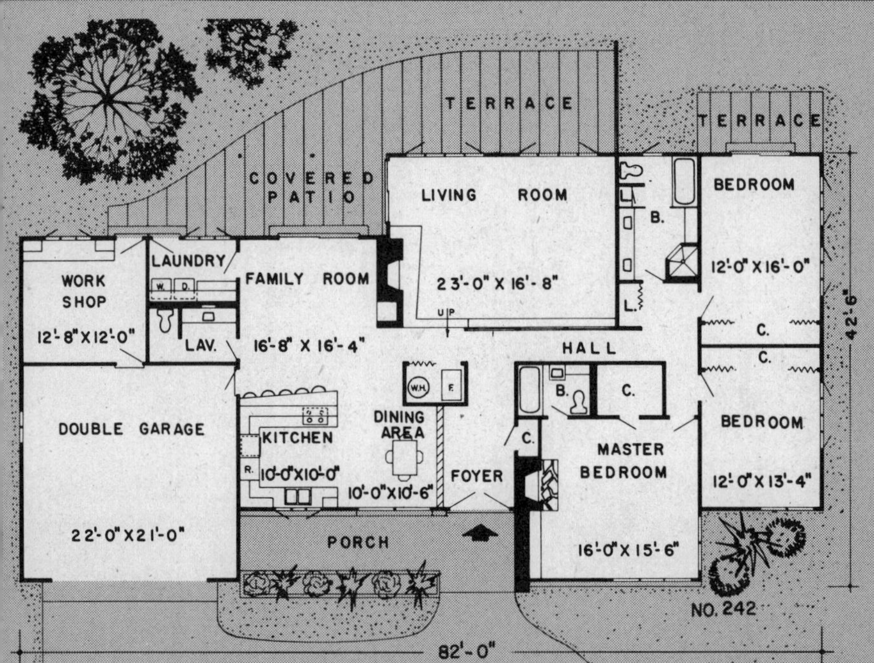

Chimney commands contemporary exterior

No. 9662—Indicative of glowing log fires within, the chimney dominates and enriches the exterior of this three bedroom contemporary home. Expanses of windows are shaded by overhangs and admit light without glaring sun. Inside, living room and family room are favored with fireplaces to spark formal or informal gatherings. Open kitchen and dining room borders a mud room that doubles as laundry, and half bath and storage room complete the area. Master bedroom is furnished with two closets and full bath, and highly efficient hall bath features double sinks, tub and shower stall.

First floor—1,903 sq. ft., Basement—1,792 sq. ft. Garage and storage—536 sq. ft.

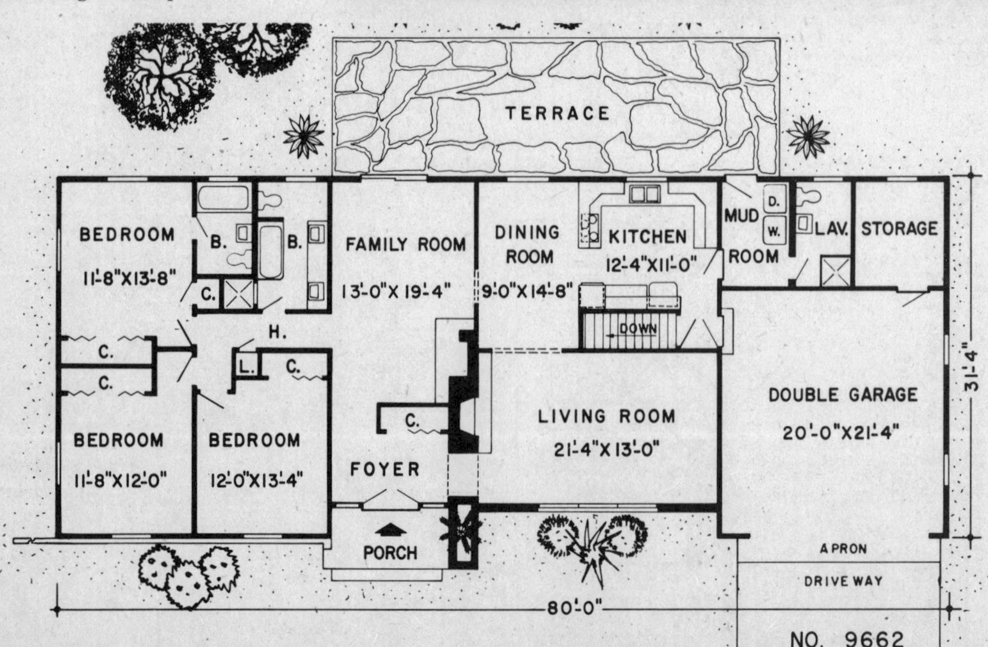

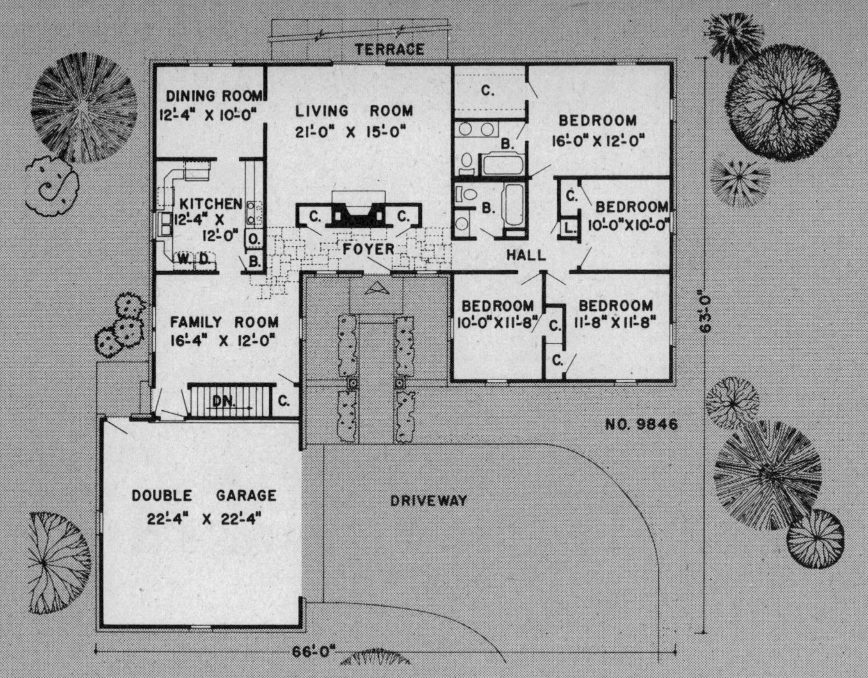

Living room encourages entertaining

No. 9846—Radiating all the charm and welcome of a French country home, this four bedroom design is highlighted by a generously proportioned living room with fireplace. Its sliding glass doors connect to the terrace and permit open, enjoyable entertaining. A formal dining room and kitchen with laundry space border the living room and the set-off family room. The master bedroom enjoys a bath and huge closet, while three more bedrooms and another bath are outlined.

First floor—2,022 sq. ft., Basement—2,022 sq. ft. Garage—576 sq. ft.

Novel design supplies sunny interior

No. 9608—Shaped like a wedge of pie, this exceptional design offers individuality and an interior bathed in natural light. Overhead clerestory windows brighten the kitchen-family room, probably the central point of the home. Three large bedrooms and two full baths edge the left side, and the master bedroom boasts its own enclosed private patio. The rear terrace is reached not only through the family room but through a bedroom and bath as well. The formal living room which accommodates a fireplace, is set to the right of the foyer and another half bath for this area adjoins the family room.

First floor—2,071 sq. ft., Garage—412 sq. ft. Storage—112 sq. ft.

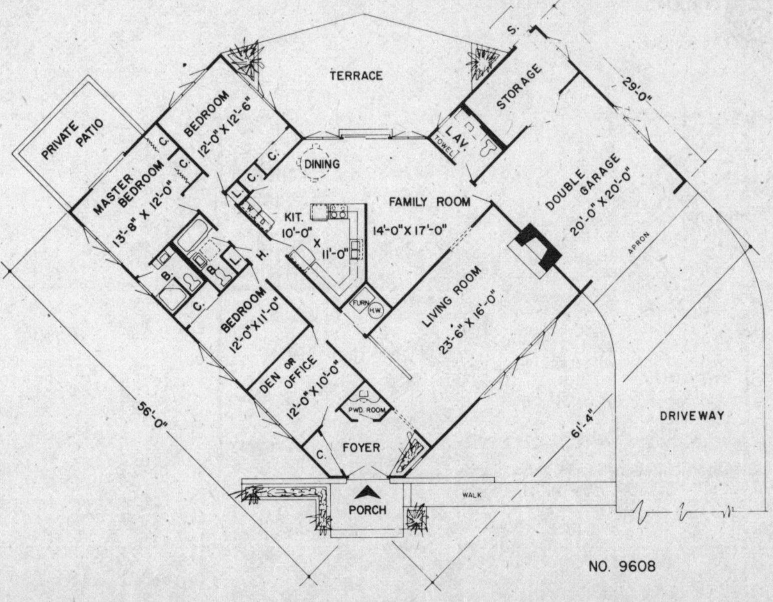

For price and order information see pages 108-109.

Multiple wing plan attired in cut stone

No. 9628—Spreading outward in three directions, this winged home is decked in vertical siding and trimmed in cupolas and cut stone. The gracious foyer admits guests to the living room or rear family room, both furnished with cut stone fireplaces. The master bedroom complex, equipped with a tub and shower bath as well as a dressing area, protrudes outward from the foyer and edges its own patio. Three more bedrooms and a bath flank the hall, which also connects to the rear terrace. A den situated behind the garage might adapt to an extra bedroom, sewing room, hobby shop, or office.

First floor—2,716 sq. ft., Basement—1,568 sq. ft. Garage—493 sq. ft.

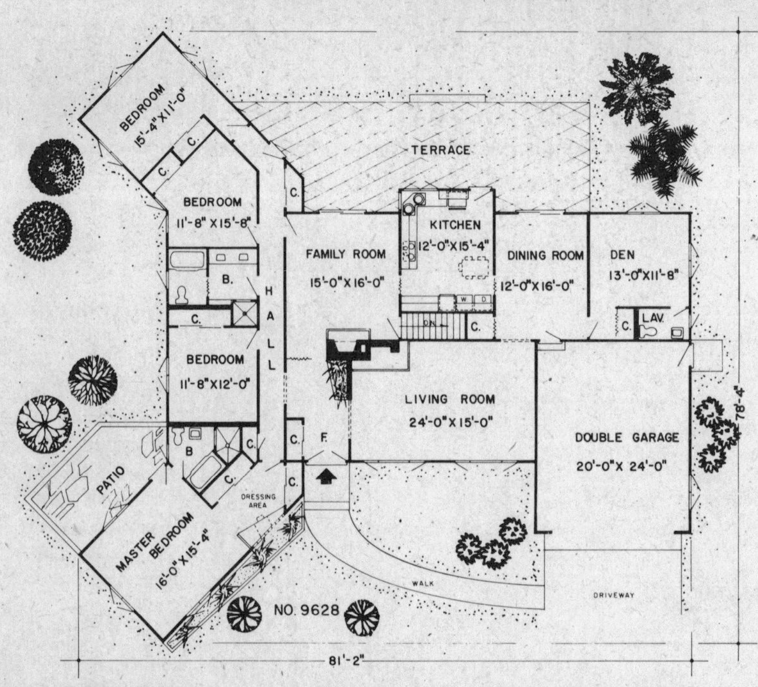

A proven plan

No. 9024—A beautiful hipped roof, brick and stone veneer ranch home. Two large bedrooms with baths, one with tub and one with shower. Each of these rooms has two large closets. There is a third bedroom and still a fourth which may be used as a den or a bedroom. The hallway leads from the front entrance back to and through the kitchen to the rear entrance. Note the large living room and the large family room, each with a fireplace. The family room is provided with folding doors to divide the room if desired and it is also provided with a half bath.

First floor—2,509 sq. ft., Basement—2,509 sq. ft. Garage—432 sq. ft.

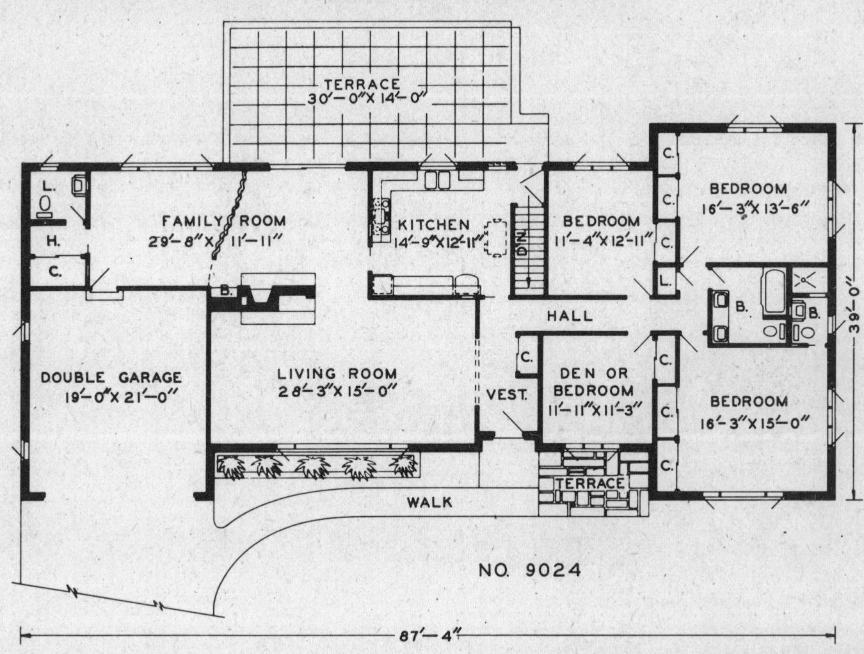

For price and order information see pages 108-109.

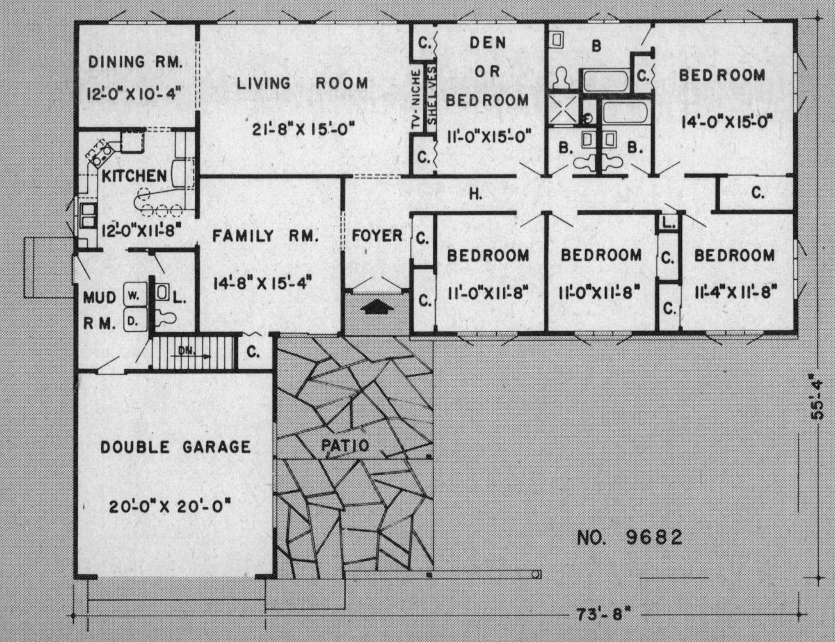

Low lines bound capacious interior

No. 9682—Intersecting brick and wood siding clothe the exterior of this low-slung ranch style, which encases a total of five bedrooms and three and one half baths. A cozy family room is favored with sliding glass doors to a flagstone patio that cascades forward on two levels. The half bath is thoughtfully placed, handy to the mud room, kitchen, and family room. Included in the bedroom wing are three full baths and five bedrooms, one of which is fringed with shelves and appropriate for use as a den or children's playroom.

First floor—2,362 sq. ft., Basement—2,362 sq. ft. Garage—420 sq. ft.

Basic ranch shows formal, informal areas

No. 19663—Careful planning and quality materials distinguish this ranch plan, which provides areas for formal and family living. Three bedrooms, each with hardwood floors, are steps from a skylit bath, and the utility room is situated off the entry, while the family room joins the kitchen.

First floor—1,516 sq. ft., Garage—364 sq. ft.

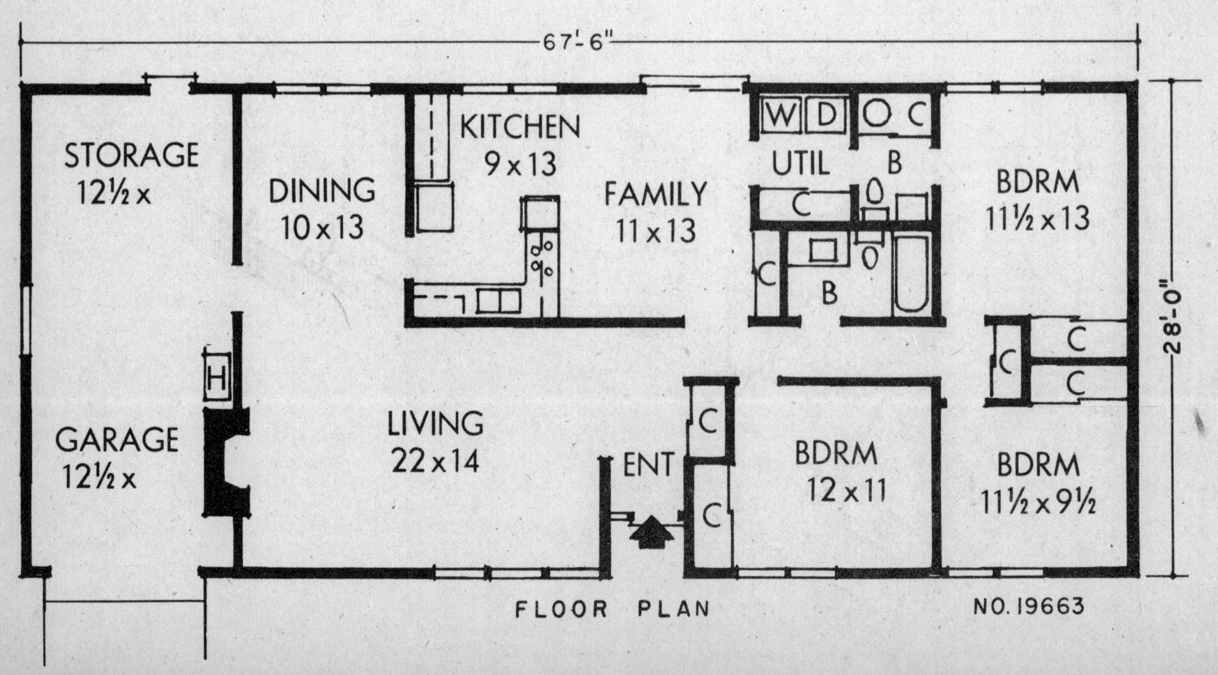

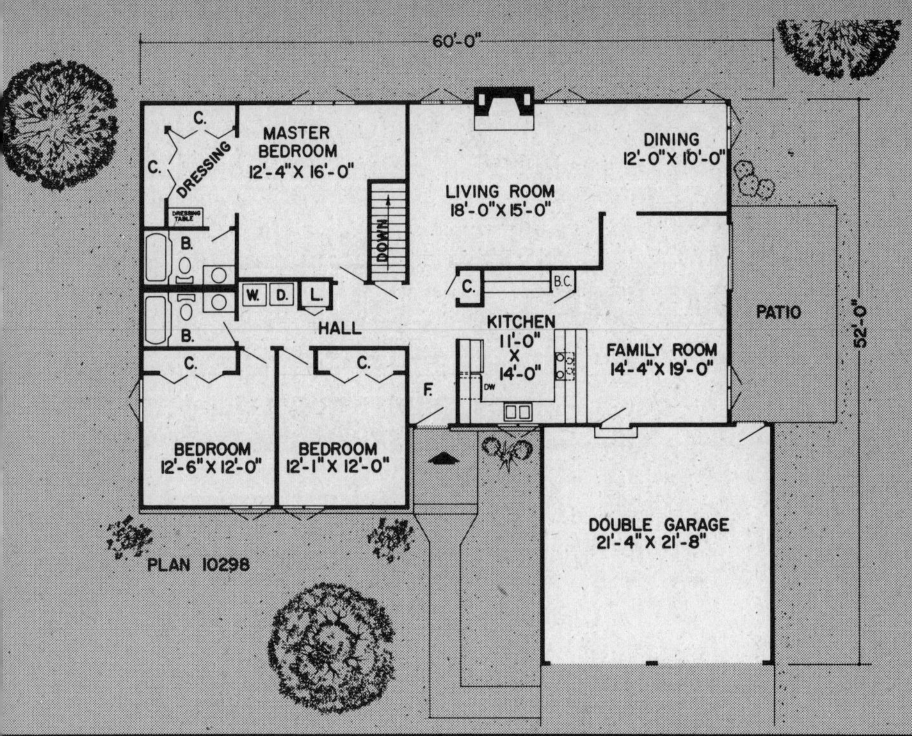

Skylights, space mark contemporary plan

No. 10298—Natural light from skylights floods this three bedroom contemporary, a well-designed home that focuses on space. Rooms are airy and include a 19-ft. family room that annexes a patio, a 12-ft. dining area, and an 18-ft. living room with wood-burning fireplace. Master bedroom is favored with a separate dressing room and private bath.

**First floor—1,889 sq. ft., Garage—491 sq. ft.
Basement—1,889 sq. ft.**

Dining area enlarged via deck

No. 19722—A spacious deck at the rear of this three bedroom design joins one bedroom and the dining area via sliding glass doors. Perfect for entertaining, the impressive living room offers a wood-burning fireplace, while the first floor study provides a quiet retreat. Two closets furnish the sizable master bedroom with adjoining bath. Plans show an opening for a spiral stairway to the balcony above the dining area.

First floor—1,640 sq. ft., Basement—466 sq. ft.

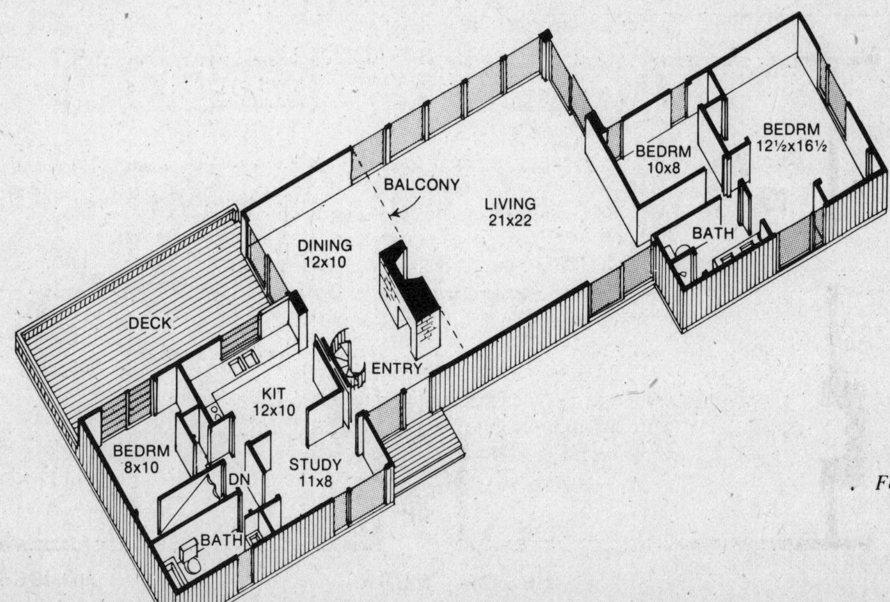

*For price and order information
see pages 108-109.*

The Product Information Source

A wealth of Building Product Information available to you free or at nominal cost. Use the order form on pages 95 & 96 to obtain the literature you want.

Appliances

ENERGY-SAVING APPLIANCES. Brochure describes energy Saver IV dishwasher line which allows consumers to turn the water heater down to 120°F and still get excellent cleaning results. Also covers time-saving trash compactors, disposers, Hot-Water dispensers, mixers and coffee mills. *Kitchen Aid Div., Hobart Corp. Price 50¢. Circle No. 402.*

KITCHEN AND BATH IDEAS. Deluxe full-color book tells how Elkay stainless steel sinks and other products give a smart contemporary look that flatters any interior . . . blends with every color scheme. New work-saving accessories in many models and sizes. *Elkay Manufacturing Co. Price 25¢. Circle No. 401.*

WASTE KING. Brochure gives specifications and photos of complete Waste King line of appliances. *Thermador Waste King, Div., of Norris Indust. Circle No. 101.*

THERMADOR. Brochure gives specifications and photos of complete Thermador line of appliances. *Thermador Waste King, Div. of Norris Indust. Circle No. 102.*

KITCHEN SINKS. Complete catalog gives specifications and photos of a complete line of stainless kitchen sinks. *Polar Ware Co. Price 25¢. Circle No. 403.*

Bathroom

SUCCESSFUL BATHROOMS. New bathroom guide offers a wealth of excellent ideas for new and remodeled bathrooms. Includes tip on bath planning, choosing good products, bathroom do-it-yourself jobs for homeowners, etc. *Heads Up Inc. Price $1.00 Circle No. 407.*

BATHE YOURSELF IN QUIET ELEGANCE. No longer is the bathroom the forgotten room in your house. We've created an atmosphere of total comfort and elegance, a pleasant place just for you . . . with softly expressed lines that are pleasing to the eye and with delicate curves that are molded for the contours of your body. Coupled with this grace and comfort is amazing practicality. *Acrylic Tubs Inc. Price 50¢. Circle No. 408.*

WHIRLPOOL BATHS. Full-color catalogue will show you the largest product line in the industry. You can choose from more sizes, shapes, colors and combination of units than anywhere else. *Jacuzzi® Whirlpool Bath. Circle No. 107.*

NEWEST THING IN ANTIQUES. Free brochure illustrates the Award Collection of high-styled faucets. The new model is an antique brass version of Delta's most popular chrome widespread. *Delta Faucet Co. Circle No. 106.*

TILE SHOWER WATERPROOFING. Asphalt membrane specially designed for waterproofing under tile shower floors. Easy to install and economical. Protects subfloors from water damage from shower. *Compotite Shower Pan. Circle No. 109.*

THE SAUNA ROOM. Brochure describes the construction and use of your own sauna. Wiring diagrams and floor plans included along with technical data on several models. *Vega Sauna Co. Circle No. 197.*

SAUNA PLANS AND EQUIPMENT SELECTION. Full-color brochure with plans and instructions for installing authentic Finnish saunas. Modular and pre-cut, sauna rooms in sizes from 4'x4' to 8'x12'. Choose one of the standard plans or follow recommendations for designing your own custom sauna. A complete line of quality heaters, controls and accessories. *Amerec Corp. Price 75¢. Circle No. 409.*

DECORATOR BATH IDEAS. Pictures a variety of decorator baths in full color. Mediterranean, traditional, contemporary, jungle and nautical designs are included. *Tub-Master Corp. Circle No. 103.*

PUSH BUTTON PLUMBING. Booklet details Ultraflo, a system which conserves both water and energy and can be installed quickly at much less cost than conventional plumbing. Reduces the water bill up to 35% and saves about 25% on the heating of water. *Ultraflo Corp. Circle No. 110.*

EVERYTHING FOR THE BATH. Full-color brochure illustrates full line of lavatories, bathtubs, water closets & whirlpools. You'll find exactly what you're looking for, with color and style to compliment your new home. *Briggs, a Jim Walter Co. Circle No. 104.*

WHIRLPOOL SPAS. Full-color catalogue shows the full range of spa's available with detailed design. *Jacuzzi® Whirlpool Bath. Circle No. 108.*

KOHLER ELEGANCE. 40 pages of colorful ideas for your bathroom, powder room and kitchen, including whirlpool baths, spas, environmental and enclosures, fiberglass shower coves, bidets, decorative faucets, cast iron kitchen sinks and much more. Booklet helps with product selection, color coordination and decorating ideas. *Kohler Co. Price $1.00 Circle No. 406.*

COMPACT PRODUCT GUIDE. A handy pocket-sized product guide on its lines of kitchen and lavatory faucets as well as its tub/shower controls is available from Bradley Corp. Faucet Div. This four color guide includes photographs and a description of the company's lines of single control faucets as well as its lines of two-handle fittings. Also covered is an explanation of Bradley's exclusive free replacement cartridge warranty. *Bradley Corp. Circle No. 105.*

JASBA-MOSAIK. The desire for perfect harmony in all items of interior home decorating is constantly growing. The development of interior decorating in recent years has more and more tended towards making bathrooms an exclusive status symbol. This beautiful full-color booklet shows new bathroom designs with the Jasbafluer ensemble. Mosaic tiles and bathroom linen in identical design and colours matching the sanitary installations. *Amsterdam Corp. Price $1.00. Circle No. 405.*

STEAMIST STEAM BATHS. Be good to your body. Experience the ultimate feeling with a Steamist steam bath. Literature contains information that shows you how to convert your shower stall or tub enclosure into a luxurious home spa with the Steamist steam bath generator. Automatic timer lets you select the time of your steam bath. Great for tired muscles, complexion, health. *Steamist Co. Inc. Price 75¢. Circle No. 404.*

WHEN ONLY THE FINEST WILL DO. A growing trend is underway that's getting people across the country into hot water . . . it's the personal spa. It's pure relaxing pleasure that once you've experienced, you'll wish you'd discovered long ago. Full-color brochure describing the ultimate in personal spas. *Avalon Spas Inc. Circle No. 111.*

BATH ACCESSORIES. Full-color folder illustrates Accentware bath accessories, as practical as they are beautiful. Includes dual-track shower bar which lets you dress your tub enclosure with draperies. Switch plates, etc., for use on any room. *Kirsch Co. Circle No. 191*

Built-ins

BUILT-IN IRONING CENTERS. Brochure illustrates a variety of built-in ironing centers ideal for homes, apartments, dormitories, home economic labs, hospitals, cabins, and hotels. The centers provide compact pressing areas without any preparation or storage problems. *Iron-A-Way, Inc. Circle No. 112.*

CENTRAL CLEANING SYSTEM IN YOUR HOME. New 32 page booklet is off the press. It provides all you need to know about installation and maintenance of a central vacuum cleaning system. Fully illustrated with drawings and photographs. *Wal-Vac Inc. Price 25¢. Circle No. 410.*

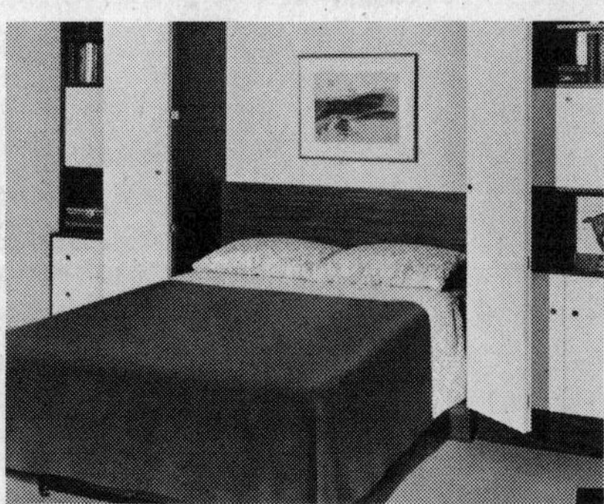

THE MASTER OF LUXURIOUS DECEPTION. Murphy Beds help to create multi-purpose rooms by providing sleeping accommodations as comfortable as the best standard bed which can also be concealed simply and safely with a minimum of effort. They are easy to conceal because they use a simple and efficient counterbalancing mechanism proven by years of field experience. Many decorating ideas and full details in brochure. *Murphy Door Bed. Co. Inc. Price 60¢. Circle 411.*

Doors & Windows

OVERLAID PLYWOOD BETTER THAN PLYWOOD. Ruf-Sawn or Stucco 316 phenolic resin overlaid plywood is available prefinished or unfinished with a wood-grain or stucco-like texture. It takes only half as much paint to cover the overlaid surface and the finish lasts up to three times longer than on raw wood surfaces. *Simpson Timber Co. Price 75¢ Circle No. 422.*

The Product Information Source

GARAGE DOORS FOR EYE APPEAL-ING STYLE. Colorful booklet shows variety of garage doors to enhance the appearance of your home. How-to instructions includes helpful hints on installing automatic door openers. *Raynor Manufacturing Co. Price $1.00. Circle No. 417.*

EASY TO INSTALL BIFOLDING DOORS. Folder illustrates five traditional bifold designs and explains why these 1 ⅜" Hemlock doors are 92% stronger than 1 ⅛" Pine bifolds. Complete installation instructions cover installing the track, adjusting doors and proper finishing. *E.A. Nord Co. Circle No. 120.*

THE WINDOW BOOK. Windows and doors and energy conservation are treated in detail in The Window Book by Fred M. Schmidt. This 136 page illustrated book identifies the ways energy is wasted and explains the specific solutions to the problems. It presents investment paybacks for these energy measures and covers many other aspects even an enlightening history of windows in the world. *Season-All Industries Inc. Price $2.95. Circle No. 416.*

WINDOWS AND CONDENSATION. Explains relative humidity, how to reduce production of excessive moisture, proper ventilation, use of vapor barriers, and warming up windows to minimize the problem. Exclusive chart identifies 200 window conditions. More than 30 colored illustrations. *Season-all Industries Inc. Price $1.95. Circle No. 415.*

ROLL SHADES & FOLDING DOORS. Literature details the beauty & durability of woven wood roll shades, as well as wood slat folding doors. All products are manufactured of choice North American Basswood. Splints are uniformly woven together using Fishermen's seine twine due to its durability. Aerolux and Temlite roll shades plus Veni-Flex folding doors are available in oil stain or enamel finishes as specified. *Aeroshade, Inc. Circle No. 123.*

VICTORIAN RESTORATIONS. 16 page four-color catalog illustrates dozens of solid wood Victorian millwork designs based upon turn-of-the century originals depicting actual applications. Entire line made from kiln dried, premium grade hardwoods. Line includes Victorian millwork for both exterior and interior use. All millwork available in any quantity. Customizing is also offered to meet specific needs. *Cumberland Woodcraft Co., Inc. Price $3.00. Circle No. 418.*

INTERNATIONAL DOORS. Brochure describes International Doors from the finest vertical grain Douglas Fir or Western Hemlock. Simpson uses original art in the designer series to create solid wood doors with accents. Traditional doors are also featured, and special care is always taken to provide elegance at a realistic price. *Simpson Timber Co. Price 75¢. Circle No. 420.*

REDWOOD PLYWOOD. Careful blend of redwood and western softwood innerplies with a selected redwood face. Offered in two basic grades and a variety of patterns. For interior or exterior use, Simpson Redwood Plywood offers a unique combination of weatherability, insulation and beauty. *Simpson Timber Co. Price 75¢. Circle No. 421.*

A POCKET GUIDE TO UNDER-STANDING CUSTOM SHUTTERS. Has just been published by OHline Corp. It contains illustrations of windows and openings showing how to select the proper shutter, measuring instructions and how to order. *OHline Corp. Circle No. 126.*

PLASTIC-VIEW. Heat reflective "See-Thru" window shades provide outstanding all weather window insulation, plus excellent protection against glare and sun damage to interior furnishings. Plastic-View shades also create daytime privacy whereby insiders can see out but outsiders can't see in . . . the security of a one-way mirror. *Plastic-View Inc. Circle No. 125.*

THE WINDOW GALLERY. Spring Crest has created a design which will capture the warmth, grace and character of your own personal style. Designed fabrics, along with the proper drapery hardware, will enhance your overall interior and when presented, will bring together the Spring Crest look throughout your entire home. *Spring Crest Co., Inc. Price $1.00. Circle No. 414.*

COMPLETE OUTSIDE BASEMENT ENTRANCE! You probably know the sales appeal direct basement access gives your home. Perm-Entry makes it easy and economical for you to include it in your home. The concrete stairwell is manufactured and installed to rigid specifications, and capped with the rugged watertight heavy gauge steel PermEntry door. *Perm Entry Co. Circle No. 122.*

ENJOY YOUR HOME MORE WITH A BILCO BASEMENT DOOR. When building your new home, be sure it has this key to a convenient, useful, safe basement . . . a modern, all-steel Bilco Door. It supplements your interior stair. Eliminates unnecessary tracking through first floor rooms. Makes storage easy, convenient. Send for free folder with sizes, construction data and names of local dealers. *The Bilco Co. Circle No. 121.*

THE COLLECTION. Beautiful, pre-engineered architectural ornamentation in lightweight modern materials. Products include: mouldings, ceiling medallions, mantels, overdoor pieces/spandrels, niches, domes, stair brackets, etc. Pre-engineered and primed for ease of installation. Great for restoration or remodeling, or for giving your home that custom-built look. All products can be painted or stained. *Focal Point Inc. Price $1.50 Circle No. 419*

WOODMASTER IS A WALL. A woodmaster space divider is your key to getting more efficient use from any interior space. Create an extra bedroom or private den at a moment's notice. Colorful brochure describes how. *Modernfold. Circle No. 119.*

DOORS OF A LIFETIME. Lifetime Doors manufactures doors of all types including flush hollow or solid core, raised panel, white pine louver/panel and more in finished or unfinished. Color brochure describes and shows all types available. *Life-Time Doors. Circle No. 116.*

STYLES IN PERFECTION. Old-world styling through modern technological procedures produces entries which reflect charm and details to enhance any entry. Fourteen styles and a multiplicity of glass options to choose from. Gallery Door Lites, the perfect choice for your next entry. *Visador Co. Circle No. 124.*

INTRODUCING WINDO-GREEN. It's an attractive way to show off plants while saving space. Windo-Green is easy to handle, easy to install and doesn't require extra structural support. For good looks, durability and quality construction Windo-Green is the only choice to make. *Acorn Building Components Inc. Circle No. 117.*

HOW TO WORK WITH WOOD MOULDINGS. For the do-it-yourselfer, this illustrated, 16-page brochure goes step-by-step through the installation of mouldings. An explanation is given of the basic types of profiles available, possible applications, and how to calculate the amount of moulding needed for a specific job. *Wood Moulding and Millwork Producers. Price 50¢. Circle No. 469.*

FOLDER SHOWS MOULDING ROLE IN AMERICAN COLONIAL. Decorating with wood mouldings to establish an authentic looking colonial design is the theme of "American Colonial". This brochure has detailed sketches showing colonial moulding profiles and illustrates their correct interior and exterior applications. *Wood Moulding and Millwork Producers. Price 50¢. Circle No. 471.*

WOOD MOULDING AND MILL-WORK. Design ideas and standard moulding profiles are depicted in the beautifully colored booklet. This 8-page brochure contains photographs of attractive room settings, highlighted with the use of wood mouldings. There are 3-dimensional drawings of standard wood moulding profiles to enable the reader to perceive their actual shapes. An excellent guide for consumers, builders, architects, designers, and do-it-yourselfers. *Wood Moulding and Millwork Producers. Price 50¢. Circle No. 472.*

WINDOW AND GLIDING DOOR AN-SWER BOOK. 24 page, full-color booklet answers the most-asked questions about windows and gliding doors. Contains special sections on window planning, energy and remodeling. Also includes full description and application photographs of beautiful Andersen wood and low maintenance Andersen vinyl-sheathed Perma-Shield windows and gliding doors. *Andersen Corp. Circle No. 115.*

RELIABLE SIMPLICITY IN DOOR CONTROLS. Reading-Dorma Closer Corp., has published a unique 16 page product guide to hydraulic door controls. This comprehensive catalog features an easy-to-read guide to ANSI numbers, and has a cross index to comparable competitive products. In addition, there is a "Typical Application" section that lists the correct product to use for a particular job application. *Reading-Dorma Closer Corp. Circle No. 114.*

BEAUTIFUL INSULATED STEEL DOORS. Detailed descriptions of the features, styles and options available with Therma-Tru steel door entry systems are contained in new catalog. Features include an improved magnetic weather strip and a redesigned, more effective door bottom and sill combination. Full-color photos of actual installations are featured. *Therma-Tru Inc. Price 25¢. Circle No. 412.*

TROCAL RIGID VINYL WINDOWS. A handsome 6-page brochure describes double-hung residential windows which combine the energy saving advantages and ruggedness of Trocal commercial windows with the flexibility required for new residential installations or remodeling. *Trocal Window Systems, Div. Dynamit Nobel of America Inc. Circle No. 113.*

DESIGN AND DECORATE WITH WOOD MOULDINGS. Contains full color photos of various household rooms showing extensive use of wood mouldings. Profile drawings of all mouldings used to create these finished designs are shown, plus suggested alternative profiles. Mouldings are a simple way to transform an ordinary room into a showplace on a minimum budget. *Wood Moulding & Millwork Producers. Price 75¢. Circle No. 457.*

FYPON ENTRANCE SYSTEMS. Brochure features molded millwork for distinctive entrance ways. Moldings, window features specialty millwork, door entrance systems, bow window roof and shutters. *Fypon Inc. Circle No. 118.*

THE FRENCH PATIO DOOR SYSTEM. Small booklet with information about Therma-Tru attractive energy-efficient and secure patio doors. They come in two styles with multiple units. Insulation facts and specifications are included in literature. *Therma-Tru Inc. Price 25¢. Circle No. 413.*

GUIDE TO ENERGY-SAVING WINDOWS. A new 16-page booklet on the subject of wood windows is now available. Topics discussed include: the components that comprise a quality window; the various styles of windows that are available; proper window locations for optimum illumination, ventilation, and fuel savings; the effect quality windows can have on energy usage; condensation; the esthetic value of windows; the care of windows and remodeling with windows. *National Woodwork Manufacturers Assoc. Circle No. 209.*

Fireplaces & Stoves

FIREPLACE FURNISHINGS GUIDE. Offers suggestions on how to decorate your fireplace distinctively with firesets, irons, spark guards, screens, etc. Hints on what is available, how to select it and where to buy it. *The Adams Co. Circle No. 129.*

PRACTICALITY & BEAUTY! Packet fully describes the best in woodburning stoves. Highly efficient as a heater using less wood for more heat. Ease of installation and many styles to choose from. *Hearthstone. Price $1.00. Circle No. 424.*

COMBINATION HEAT. Brochure describes free-standing wood-burner that uses wasted heat from the bottom and back of the stove and forces the heated air along the floor for maximum efficiency and heat distribution while still giving the benefit of radiant heat from the stoves heavy metal surfaces. *Big Timer Wood Stoves Inc. Circle No. 132.*

WOOD-BURNING & GAS STOVES. This brochure presents photos and descriptive information on the complete line of U.S. Stoves. Includes what to look for in a wood stove, results of efficiency tests and heating large areas most effectively. *United States Stove Co. Circle No. 131.*

BUILT-IN FIREPLACES. Fireplaces by Martin offer a complete line of zero-clearance, build-in anywhere, manufactured fireplaces. Available in 28", 36", and 42" sizes. All units are UL listed. Full-color brochure available. *Martin Industries. Price 25¢. Circle No. 427.*

QUADRA-THERM FIREPLACES. The Quadra-Therm is a true heat-circulating fireplace, taking room air in and around the enclosed firebox and returning it, heated, to the home. Energy-saving glass doors complete the efficient package. Full color brochure available. *Martin Industries. Price 25¢. Circle No. 426.*

EASY LOG LIGHTING. Blue Flame log lighting system makes heating with modern pre-fab fireplaces more energy efficient by using less than 1¢ of natural gas to quickly ignite a roaring log or coal blaze. System is easily installed in either pre-fab or existing masonry fireplaces. Saves oil, gas and electricity in home heating. *Canterbury Enterprises. Circle No. 133.*

WOOD MANTEL PIECES. A 10-page catalog featuring more than thirty handcrafted wood mantels, many in room settings. A series of "before" and "after" photos illustrate how proper accenting with wood mantels and decorator panels can transform a cold, uninteresting wall into a warm, inviting accent feature in any home. Brochure includes techniques in proper measurement to assure fast and accurate mantel installation. *Readybuilt Products Co. Price $2.00. Circle No. 458.*

WOOD SPLITTING MADE EASIER.
Literature tells of unique new axe which splits many woods in a single stroke. Chopper 1™ possesses rotating levers which transform the downward stroke to a powerful outward force and also prevents the axe from sticking in the wood making splitting faster and easier than ever before. *Chopper Industries. Circle No. 128.*

FIREPLACES AND MORE FIRE-PLACES.
Literature displays full line of fireplaces. There's one to suit every decor from the ultra modern to the rustic early American. Complete with energy saving and decorating ideas. *Malm Fireplaces Inc. Circle No. 127.*

FIREPLACE STOVES.
Full color detailed brochure tells about discovering heating with wood and coal in comfort and style with a multi-fuel Comforter Stove. Quality, efficiency, and beautiful fireviewing are the outstanding features of this classic 24-hour burning cast-iron heater. *Comforter Stove Works. Price $1.00. Circle No. 476.*

OCTA-THERM FIREPLACES.
Designed for zero-clearance installation, the Octa-Therm offers virtually unlimited installation flexibility. The Octa-Therm is capable of ducting heated air into as many as three separate rooms for real supplemental heating. Full-color brochure. *Martin Industries. Price 25¢. Circle No. 425.*

LET'S TALK ABOUT FIREPLACES.
Discussion by Richard Le Droff, master chimney mason, on fireplaces, selection, style and design, flue and chimney location problems of safety and technical problems of construction. 8 page full-color brochure illustrates many units and possibilities of more than 1,000 different fireplaces. *The Adams Co. Circle No. 130.*

FORGET HAVING YOUR WOOD FIRE DIE OVERNIGHT.
The Woodburner is a publication produced by the All Nighter Stove Works, Inc., which describes the latest in long burning wood and coal stoves. Packed with helpful ideas ranging from how to decide on the proper stove to how to pick the best wood for your new stove while saving over 40% of your fuel bill each winter. *All Nighter Stove Works, Inc. Price $1.00. Circle No. 423.*

FIREPLACE INSERTS.
The latest fireplace inserts deliver both high heat output and high efficiency. Unique design offers see-through glass doors for viewing the beauty of the fire — and interchangeable air-tight doors for extended burn times. Product is UL tested and listed. Includes dual variable-speed internal blowers for flush-face installation and stainless steel heat exchanger for long life. *Thermograte Inc. Circle No. 216.*

Floors & Walls

VALLEY FORGE PRE-FINISHED PLANK.
The first of its kind, Valley Forge offers the customer not only color but wood species never before available in a factory pre-finished plank. Wood species are red oak, white oak, angelique teak. All species are pre-finished with two coats of Dura Seal and one coat of hot melt wax. All are available with walnut plugs inserted or without plugs. *Harris Manufacturing Co. Price $1.00 Circle No. 479.*

WALLCOVERINGS.
New full-color 4 page brochure on "Donghia" collection of wallcoverings and companion fabrics now available from James Seeman Studios. *James Seeman Studios, Div. of Masonite Corp. Circle No. 140.*

TRADITIONALLY SPEAKING.
A new, 4-page, full-color and fully illustrated brochure on the "Traditionally Speaking . . ." collection of wallcoverings & companion fabrics, is now available from *James Seeman Studios, Division of Masonite Corp. Circle No. 183.*

INTERPRETATIONS.
A new, 4-page, full-color and fully illustrated brochure on the "Interpretations" collection of gravure-printed, fabric-backed, pretrimmed, strippable, washable/scrubbable wallcoverings, is now available. *James Seeman Studios, Division of Masonite Corp. Circle No. 184.*

DECORATIVE STONE.
Brochure describes manufactured building stone veneer made from lightweight concrete. Although available at a fraction of the cost of natural stone, it is hard to distinguish from the real thing. The light weight permits easy installation over nearly any existing or new surface. *Eldorada Stone Corp. Circle No. 138.*

GENUINE HARDWOOD FLOORS.
Full-color brochures details beautiful Applachian Oak parquet flooring. Elegance has never been so practical. *Pennwood. Circle No. 145.*

CEILINGS ARE LOOKING UP.
Full-color brochure illustrates a collection of decorative beams that are easy to install and will add that touch of distinction to your home. *Laminated Timbers Inc. Circle No. 141.*

THE PETITE LOOK.
Booklet describes the perfect mix . . . small prints and smart ideas for bedroom, dining and living room, and kitchen. Wall covering at it's best. *Panta Astor Wallcovering. Circle No. 135.*

THE PREFERRED LOOK.
The newest fashion trend today is pattern and you can create a distinctive look of pattern in your home with this collection of easy, inexpensive decorating ideas for bedroom, dining, living room and kitchen. *Panta Astor Wallcovering. Circle No. 134.*

SHINE-EASE NO-WAX FLOOR.
Full-color booklet describes the ease of installing and maintaining a beautiful floor with the use of Azrock floor tile. Shine-ease will offer you easy living extras at a most affordable price. *Azrock Floor Products. Circle No. 143.*

ALTERNATIVE FLOORING.
Brochure illustrates the use of ceramic floor tile. Tile that is durable enough for floors and yet light enough to use with coordinating walls. *Marazzi USA Inc. Circle No. 139.*

NATURESCAPES.
Brochure offers an exciting design alternative for residential or commercial decor. Naturescapes photomurals are durable, dry-strippable, and meet all institutional standards. *Naturescapes, Inc. Price $1.00. Circle No. 431.*

NATURESCAPES.
The collection consists of works by the finest photographer/naturalists. Reproduced on the most stable grade synthetic available. For a spectacular presentation of these beautiful images in spreads that are 3 feet across, without the distraction of roomsets order the hard cover Naturescapes Design Book. *Naturescapes, Inc. Price $20.00. Circle No. 430.*

NEW LOOK. Brochure describes how you can give your home a "New Look" with lustrous hardwood parquet floors. There is an extra special warmth and cordial feeling you get with parquet floors. Your parquet floors will be a conversation piece and the envy of those around you. *Peace Flooring Co., Inc. Circle No. 137.*

ULTIMATE IN HARDWOOD FLOORING. Brochure presents the complete product line, from the custom classics and end grain floors to basic plank and parquet. Line is directed toward the upper-end specifier and consumer for that special residential accent area. *Kentucky Wood Floors Inc. Price $1.00. Circle No. 428.*

CHICKASAW HARDWOOD FLOORS, WALLS. Color photos and a complete description of the wide variety of planks, strip and parquet floors from Chickasaw plus unique Pecan Wall Plank. *Memphis Hardwood Flooring Co. Price 50¢ Circle No. 481.*

THE PANELING BOOK. Paneling installation can be a snap, even in problem areas with the help of Georgia-Pacific's colorful new 28-page booklet, The Paneling Book. How-to instructions, diagrams, and full-color photos of rooms you'll want to duplicate. Make this your complete guide to paneling any room, whether building or remodeling your home. *Georgia-Pacific. Price $1.00 Circle No. 429.*

KITCHENS AND BATHS. Full color booklet packed with loads of decorating ideas for your kitchen and bath. Shows the beauty of wall coverings and how easy it is to install. *Astor Wallcovering by Pantasote. Circle No. 136.*

CLASSIC CHAIRS - & MORE - BY W&Z. Color brochure illustrates a variety of imported chairs and tables, both contemporary and traditional, of wood, chrome, leather, wicker, marble and glass. *Walker & Zanger, Inc. Price $2.00 Circle No. 480.*

SOLID OAK PARQUET FLOORING. Literature details Hartco, the world's largest manufacturer of solid oak parquet flooring, offering parquet in smooth and textured finishes for residential and light commercial jobs. Hartco Impregnated parquet is also available for heavy-traffic commercial installations. To complete its flooring "system", Hartco offers adhesives, prefinished moldings, and floor care products. *Tibbals Flooring Co. Circle No. 206.*

Furniture & Furnishing

FURNITURE. Now you have finished your home you want furniture to complete the picture. Catalog gives hundreds of decorating ideas and shows the complete line of furnishings for every room in your home. *Terra Furniture Inc. Price $3.00. Circle No. 432.*

TABLE TOPS. Literature shows table tops that are geared for end use such as dining room tables, dinettes, coffee tables, end tables, or counter tops where the physical properties of our material, which is called Radwood, can offer styling and design ideal for the residential or contract market. *Radiation Technology, Inc. Circle No. 146.*

SHELVING. Series of full-color folders explains and shows three shelving styles, one wall-hung and two freestanding twist togethers. Special units for stereo equipment. Designed and finished to look like expensive furniture, but the cost is less. *Kirsch Co. Circle No. 192.*

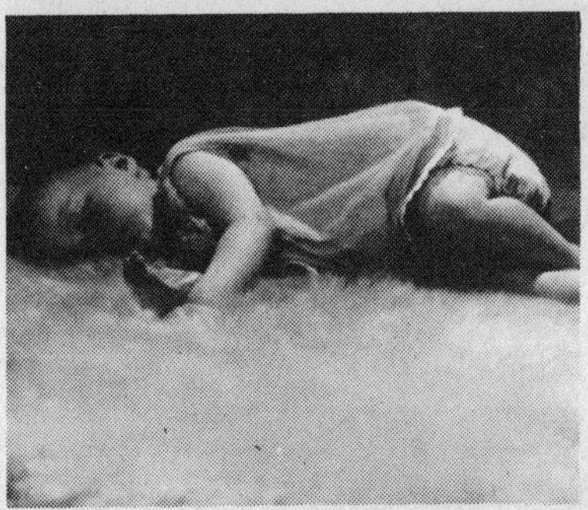

ELEGANT SHEEPSKIN & LAMBSKIN RUGS. This color brochure describes the beauty, warmth and durability of a variety of luxuriously soft wool, sheep and lambskin rugs, the ultimate in dynamic decorating. A special offer on these exceptional New Zealand woolskins is made possible through a limited shipment. New Zealand produces some of the highest quality wool anywhere due to the moderate climate. In most wool producing regions of the world, the cold winters will produce brittle areas in the wool fiber, causing these fibers to gradually break and making the fleece thinner. This doesn't happen to New Zealand wool. New Zealand care has converted one of nature's most versatile resources into fine prestigious furnishing rugs to add distinction and elegance to any home. Find out how you, too, can own one of these most delightful decorator pieces. *Treasures. Circle No. 210.*

Hardware

DECORATIVE HARDWARE. 32 page decorating booklet shows how to select and install hardware items to coordinate each room with matching style door locks, bath hardware, cabinet hardware, and wall switch plates. Several designer hardware styles are detailed. *Amerock Corp. Price 25¢. Circle No. 433.*

HANDCRAFTED HARDWARE OF THE COLONIES. For over a third of a century, Acorn has been reproducing the charm and warmth of Early American Hardware. Booklet shows full line of antique looking hardware for your home. Beautiful yet sturdy and practical. *Acorn Manufacturing Co. Inc. Price $2.00. Circle No. 434.*

CERAMIC CLIP-ON BATH ACCESSORIES. Brochure describes easily installed bath accessories, using either clip or fastick backplate, by the do-it-yourself consumer. Available with either ceramic or wood backplate. Ideal for use in old and new bathrooms. Ceramic knobs and drawer pulls are also available. *Lenape Products, Inc. Price $1.00. Circle No. 467.*

CLOCKS FOR ANYWHERE. New from Howard Miller Clock Company is a 68-page catalog outstanding in its illustrative and descriptive presentation of the manufacturer's extensive collections of traditional clocks. The catalog is visually dramatic, and unique in its techniques of merchandising floor, wall and mantel clocks and quartz alarms. *Howard Miller Clock Co. Price $2.00. Circle No. 468.*

Heating & Cooling

ENERGY SAVINGS WITH NATURAL VENTILATION. Bulletin describes how whole-house ventilation system can be used for summer comfort and energy savings by reducing air conditioning. Details advantages and describes the solid state speed control and timer control features. Installs in any home. *Kool-O-Matic Corp. Price 25¢. Circle No. 435.*

THE ENERGY SAVERS. Brochure tells of energy saving products designed to cut down on high utility bills. Everything from air deflectors to insulation. *Deflecto Corp. Circle No. 149.*

HOME COMFORT VENTILATION. Brochure covers entire line of energy saving ventilating fans. Product line includes whole house attic fans and roof mount power ventilators. *Attic Aire Division of Air America Corp. Circle No. 215.*

SOFTHEAT. Booklet describes hot-water electric baseboard units that are specially engineered to warm the cold air that flows from windows and outside walls. Utilizes natural air currents to warm the room gently and thoroughly. *Intertherm Inc. Circle No. 148.*

HUMIDIFICATION. Booklet discusses relative humidity and lists several points to consider when purchasing a humidifier. *Research Products Corp. Circle No. 151.*

VENTILATING AND CIRCULATING FANS.
20 page brochure covers the entire line of ventilating and circulating fans for residential, commercial, and industrial uses, with application photos. Plenty of data on accessories, specs, installation tips and warranty information. *Comfort Conditioning Division/Robbins & Myers, Inc. Circle No. 153.*

THE HOT SHOT.
Carrier's heat reclaim device for domestic hot water systems, automatically recycles heat energy from a residential air conditioner or heat pump system to a home's hot water system. Results in reduced domestic hot water energy consumption costs. Literature describes concept, operation, dimensions and connections, and typical piping and wiring arrangements. *Carrier Corp. Circle No. 195.*

CEILING FAN.
16 page full-color brochure relates the history of the old-fashioned paddle-blade Hunter ceiling fan and why it is making a big comeback as an energy-saver. Includes installation tips. *Comfort Conditioning Division/Robbins & Meyers Inc. Price $1.00. Circle No. 436.*

DELUXE THROUGHOUT.
Brochure features range hoods with classic styling showing a "Black Glass" control panel. Satin smooth baked enamel finish is available in appliance matching colors. Convenient solid state control for powerful dual centrifugal blower moving air efficiently through improved filtering system. *Broan Mfg. Co. Inc. Circle No. 186.*

FAN FORCED WALL HEATER.
Literature describes economical heat where you want it, when you want it. Fast-heating element provides 1430 watts of fan-forced heat in just 10 seconds for instant bathroom comfort. Saves energy — no need to turn up the central thermostat. *Broan Mfg. Co., Inc. Circle No. 187.*

CARRIER'S WEATHERMASTER III.
Literature describes the only heat pump system on the market which houses the compressor and all critical controls out of the weather. Compressor and controls are installed in garage, utility room or basement, meaning improved system reliability as critical components are protected from rain, snow and freezing temperatures. Concept, system operation, features, accessories, physical data and dimensions, and performance data all included. *Carrier Corp. Circle No. 194.*

SOLARMATE HOT WATER SYSTEMS.
6-page brochure describes the quality line of efficient and dependable solar hot water systems. *Lennox Indust. Inc. Circle No. 150.*

CARRIER.
The Carrier line of upflow gas-fired furnaces, featuring solid-state controls, low noise levels and high operating efficiency, consists of both standing pilot (58GS) and electronic ignition (58GP) models. Literature describes full line, options (including flue damper), solid-state control systems, features, accessories, physical dimensions, selection procedures, performance and application data, also depicts typical installations. *Carrier Corp. Circle No. 196.*

ALPHA HEATING SYSTEMS®.
Combination infra-red and forced air electric heating systems offer a highly efficient method of heating all or parts of homes, apartments, or, offices. Alpha I is a portable piece of furniture and the Alpha II, III are in-wall units used in new and retrofit construction. Easy to install and U.L. approved. Some call Alpha, "Solar in a Box". *Energy Dynamics International. Circle No. 207.*

AIR CLEANING.
Booklet describes air cleaning and presents efficiency data on common household pollutants and how to control them with a new type air cleaner. *Research Products Corp. Circle No. 152.*

Kitchens & Cabinets

DISTINCTIVE CABINETRY.
Full-color brochure shows custom cabinetry, storage, convenience and decorator features for kitchen and bath. *J.B.'s Cabinet and Appliance Center. Circle No. 159.*

LITTON FULL-LINE BROCHURE.
Four-color brochure gives description, specifications and photographs of Litton's full line of microwave ovens and ranges. *Litton. Circle No. 214.*

EXCLUSIVE KITCHENS.
This four-page color brochure tells you how to color coordinate and decorate your kitchen. Exclusive electric decorator series allows you to color coordinate your kitchen in 999 electrifying styles. See the unlimited design possibilities in kitchen cabinets. *Rich Maid Kitchens, Inc. Price 30¢ Circle No. 440.*

KITCHEN CABINETS.
Wood-Hu Kitchens is a custom manufacturer of all wood kitchen cabinets. Color brochure describes variety of cabinets all designed with todays fashion in mind. Showing 32 different door styles. *Wood-Hu Kitchens, Inc. Price 50¢. Circle No. 443.*

QUALITY CUSTOM KITCHEN CABINETS.
Literature with color photographs of custom cabinetry and door styles with unique styling and durable construction. They come in walnut, cherry, oak, birch, mahogany/cherry, walnut/oak and other natural wood finishes. New styling and accessory cabinet features. *Connor Forest Industries. Circle No. 158.*

LIVING CENTER.
Distinctive kitchens designed for the living patterns of your family are shown in full-color booklet. Planning procedures to use space wisely for well-organized step saving work areas are discussed thoroughly. *Kitchen Kompact Inc. Price $3.00. Circle No. 445.*

THE CENTER OF FAMILY LIVING.
New four color packet, while showing cabinet styles of oak, pine, pecan and maple, provides necessary and helpful information on kitchen design and layout as well as decorating ideas. Matching bathroom cabinetry, hutches and wall systems provide added storage ideas for every room of your home. *AristOKraft. Price $1.00. Circle No. 444.*

CABINETS.
Beautifully designed contemporary door styles accent today's modern kitchens. Birchcraft's fine assortment of stylish patterns allows you to choose the perfect door to fit your personal tastes as well as the wood and stain finish that best compliments your overall design. Ranging from the sleek, ultra-modern to country casual, the Birchcraft line of contemporary door styles is sure to compliment the most epicurean tastes. *Birchcraft Kitchens. Circle No. 154.*

NEWEST IN STORAGE.
Literature describes a unique new furniture storage system that is scheduled to be offered nationally when the franchise network is completed. The system is designed and manufactured in Europe, where storage space is a real problem. It is iP20's simplicity and flexibility that makes it so attractive. The shelves, drawers, cabinets and accessories are easily set up with only a screwdriver. *iP20. Circle No. 156.*

HALL MACK BATH ACCESSORIES.
Full-color catalog of finest bath accessory lines for remodeling and building. Complete lines offer as many as 25 accessories in each different finish. Solid brass Aristochrome included. Woods combined with brass. Built-in bath scales to grab bars. *Nutone Housing Group. Price $1.00 Circle No. 439.*

AJAX CABINET ORGANIZERS. Expand storage space conveniently in kitchens, baths, closets, playrooms and workshops. Rev-A-Shelves come in D-shape, full circle, pie-cut and kidney shapes. Swing-out units, door storage bins and single Rev-A-Trays complete the line. *NuTone Housing Group. Price $1.00. Circle No. 438.*

KING MINI-KITCHENS. A complete line of mini-kitchens from 30 to 72 inches wide, for a total kitchen where only limited space is available. All Mini-Kitchens have a refrigerator, a gas or electric range, a stainless steel sink and a seamless, stainless steel countertop. Units 48-inches or more also have a gas or an electric oven. Optional accessories include wall cabinets, waste disposer, range hoods and light fixtures. *King Refrigerator Corp. Price 25¢ Circle No. 441.*

ORIGINAL EUROPEAN STYLE KITCHENS. A 160 page color catalog of modern kitchens, cabinets, and accessory items from German's largest kitchen manufacturer. Now available in the U.S.A. these original European designed kitchens have a 5 year limited warranty (guarantee) *Alno Kitchen Cabinets, Inc. Price $3.00. Circle No. 463.*

KITCHEN & BATH PLANNING GUIDE. This 16-page color booklet contains helpful suggestions on how homeowners can get the "dream kitchen" or bath of their choice. It covers subjects such as efficient "work triangles," where to get professional help, and a consumer guide to cabinet selection. *National Kitchen Cabinet Assoc. Price 35¢. Circle No. 478.*

SPECIFICATIONS AND ACCESSORIES GUIDE. The literature contains diagrams and dimensions for all units, and shows Haas' wide array of accessories and specialty cabinets. Complete specifications and a color page of the full line of storage accessories are also included. *Haas Cabinet Co. Circle No. 208.*

DISTINCTIVE CABINETRY. Packet of full-color literature showing custom cabinetry, storage, convenience and decorator features for kitchen and bath. *Quaker Maid. Price $2.50. Circle No. 442.*

MERILLAT CABINETRY. Beautiful 12 page, 4 color brochure describes all of Merillat's cabinet lines. Each two page spread features a decorator styled kitchen which offers a multitude of ideas for the homeowner. *Merillat Indust. Inc. Price $1.00. Circle No. 464.*

Lighting

A TOUCH OF CLASS. Liviton's Decora line of rocket switches, available in all home harmonizing colors are indeed a decorative touch in a home that adds "class". Dimmers both of the wall variety and table top are also important features. Even dimmers that are touch sensitive, has a memory and a true decorator look. *Leviton Manufacturing Co., Inc. Circle No. 162.*

ARCHITECTURAL LANTERNS. Trimble House architectural lanterns feature colonial, victorian and contemporary styling to harmonize perfectly with buildings and landscape new and old. All lanterns are totally cast in aluminum offering maintenance free, functional construction with enduring elegance of design. *Trimblehouse Corp. Circle No. 161.*

HOME CONTROL SYSTEM. From Leviton, a modular remote control plug in system which permits the user to control up to 16 lights or appliances from one location in addition to convenience a dimming feature provides energy savings. A single button for all lights, off or on, also has the advantage of security. Starter kit includes control box and 2 modules. *Leviton Mfg. Co., Inc. Circle No. 213.*

LIGHTING. 120 page catalog of lighting fixtures for home building and remodeling. Includes Sterling recessed and surface-mounted incandescents and flourescents for every purpose. Complete section on Trackline's one and three circuit systems with wide choice of handsome shields. Energy saving fixtures emphasized. *NuTone Housing Group. Price $2.00. Circle No. 437.*

NATURAL SKYLIGHTS. Send for this full-color brochure describing standard size sky-light models. Includes photographs of room interiors and the names of dealers in your area. *Naturalite, Inc. Circle No. 160.*

Patio & Yard

ORNAMENTAL GATES & FENCES. Color catalog illustrates a selection of orna-

mental gates and panels for fencing. Master plan illustrates how to design a brick wall containing ornamental fence panels. Catalog also includes complete line of outdoor furniture for poolside, beautiful fountains, urns, planters, garden plaques and mailboxes as well as a variety of small gift items. *Moultrie Mfg. Co. Price $1.00. Circle No. 447.*

OLD SOUTH REPRODUCTIONS OFFERS A RETURN TO ELEGANCE. Classic and timely styling from bygone days, masterfully recreated for today. Cast aluminum tables, chairs, settees, coat rack, umbrella stands . . . hundreds of singularly distinctive pieces for home and patio. Full-color catalog depicts wide range of items in lovely finishes. *Moultrie Mfg. Co. Price $1.00. Circle No. 460.*

CREATIONS OF AGELESS GRACE AND BEAUTY. Fine reproductions of classic designs in accents and decorator pieces are lovely and lasting in cast aluminum. Color catalog shows items in actual Southern homes. Decorative ideas for gardens, lawns and patios. Hundreds of ideas to give or to get. *Moultrie Mfg. Co. Price $1.00. Circle No. 461.*

RETURN TO ELEGANCE. An exclusive and unique collection of accents, accessories, and furniture, reproduced from quaint and stately patterns of yesteryear are pictured in full color. Cast in durable aluminum are fountains, urns, mirrors, planters and plant stands, tables and chairs, all in wide choice of color finishes. *Moultrie Mfg. Co. Price $1.00. Circle No. 462.*

REDWOOD, HOT WATER & YOU: A PERFECT MIXTURE. A colorful brochure describing the joys of owning your own hot tub. Gordon & Grant redwood hot tubs are individually hand-crafted by highly skilled coopers from carefully selected clear all-heart kiln-dried vertical grain redwood. Every tub is an individual. *Gordon & Grant. Circle No. 189.*

DECKS ADD FUN & VALUE. This year enjoy your home inside and out by investing in contemporary outdoor living. Redwood Design-a-Deck Plans Kit can help you get started. As with any project, careful advance planning assures more efficient deck construction. This kit helps minimize confusion by giving you everything you need to know, including materials lists, information on which redwood grades to use and where, which finishes to apply and how. Nails and fastening systems, as well as precut deck patterns, 20-page instruction man-

ual and planning grid. *California Redwood Assoc. Price $4.00. Circle No. 473.*

FOUNTAINS & WATERFALLS. Full-color literature shows the beauty of nature's sparkling streams and the serenity of her still waters in lightweight fiberglass creations. Easy to install and affordable. These products will transform their surroundings into fascinating garden areas, places to relax and escape from the tensions of the day. *Hermitage Gardens. Price $1.00. Circle No. 446.*

SHOWERS OF DIAMONDS. Fountains by Rain Jet. Sculptured patterns from rotating fountain head provides arresting beauty. Alive! Vital! Fascinating! Light and color blender available for jewel-like effects. Water recirculates. Complete fountain bowls to 8 feet. Full color catalog. *Rain Jet Corp. Price 50¢. Circle No. 465.*

UNDERGROUND SPRINKLERS. Brochure shows how to install Rain Jet permanent underground sprinkler system with a minimum of sprinkler heads and digging. Patented nozzles distribute rain-like droplets evenly. Choice of heads includes squares, strips, circles, etc. Flexible pipe and full flow fittings make installation easy. *Rain Jet Corporation. Price 50¢. Circle No. 466.*

GARDEN SETTINGS. New eight-page color booklet, Redwood Garden Settings for Spas, Tubs, Pools, features spectacular idea-starting hot tub surroundings, poolside decks and spa shelters, all using the economical redwood garden grades. Design tips can help you build a luxury redwood retreat of your own. *California Redwood Assoc. Price 50¢. Circle No. 475.*

Roofing

REMODELING IDEAS. New color brochure packed with ideas on how to remodel with red cedar shingles and handsplit shakes. Photographs show remarkable step-by-step transformation of homes. Projects from simple over-roofing to major remodeling. *Red Cedar Shingle & Handsplit Shake Bureau. Price 25¢. Circle No. 450.*

SUPERIOR INTERIORS. Handsome, new full-color brochure showing unique decorating ideas using red cedar shingles or handsplit shakes on interiors. See how these products bring fascinating texture and color into a room. "How to apply" information shown. *Red Cedar Shingle & Handsplit Shake Bureau. Price 35¢. Circle No. 451.*

HOW TO DO IT . . . AND SAVE. Color brochure featuring economy grade red cedar shingles and handsplit shakes. Filled with ideas on the use of #2, #3 and #4 shingles plus #2 shakes on walls, planters, furniture and in many other ways around the home. *Red Cedar Shingle & Handsplit Shake Bureau. Price 10¢. Circle No. 449.*

OVER-ROOF WITH CEDAR. Color brochure showing how home owners can save applying a beautiful cedar shingle or shake roof right over their old roof. Detailed "how to" information plus data on tools and techniques. *Red Cedar Shingle & Handsplit Shake Bureau. Price 10¢. Circle No. 448.*

Solar

SOLAR HEATING. Literature describes a series of components for solar and radiant heating that are highly durable and low in cost. The versatile, flexible material of SolaRoll components makes the systems ideal for retrofit applications, and they integrate neatly into new construction. In addition they have developed a whole new solar technology which is extremely easy to install. *SolaRoll, Division of Bio-Energy Systems, Inc. Circle No. 170.*

SOLAR HOME SYSTEMS. Brochure describes the workings of Solar Energy and how you can use it to heat your home in winter and cool it in summer. *Solara Technical Center. Circle No. 169.*

HEAT WITH THE SUN! Save our vital resources while you save on your monthly utility bills. Brochure describes solar collectors that heat your home and water. *Sunflower Energy Works. Circle No. 168.*

LET THE SUN WORK FOR YOU. Literature describes complete line of solar heating equipment for domestic, commercial and industrial applications, consisting of solar collectors, universal mounting frame and automatic temperature control system. *Solar-Eye Products, Inc. Circle No. 185.*

HOT WATER THE SOLAR WAY. Full-color brochure describes economical approach to domestic hot water also shows pre-plumbed tanks that are used with the collector to form the main components of a Solar D.H.W. System. *Solar Warehouse, Distributors for Jet Air Solar Products. Circle No. 167.*

TODAY'S PRACTICAL, ECONOMICAL ANSWER TO HIGH HEATING COSTS. Brochure describes air-to-air active system which functions in the flat plate category but is not, in reality, a flat plate collector. Easy and economical to install. *Deltair Solar Systems. Circle No. 165.*

SOLAR COLLECTOR INCREASES BTU EFFICIENCY. Brochure describes breakthrough in solar collector design and manufacture enabling a significant boost in BTU's per dollar. Two exceptionally light models are available including a double glazed acrylic-Teflon collector. *Sunearth Solar Products Corp. Circle No. 164.*

SOLAR COLLECTOR PANELS. Flat plate collector panels for solar systems to heat building interiors and domestic hot water supplies are described in brochure. *National Solar Supply. Circle No. 163.*

THE ACROSUN COLLECTOR. The AcroSun collector is a well constructed all copper collector with glass plate, aluminum annodized housing, aluminum backing and is fully insulated. It is available in Flat Black Paint or Black Chrome. *AcroSun Industries Inc. Circle No. 182.*

AMETEK POWER SYSTEMS GROUP. Currently offering its Sunjammer solar collectors to qualified contractors and builders for installation in residential space and hot water heating systems. The product incorporates an aluminum extruded housing finished with the Sherwin-Williams Power-Clad® baked enamel paint finish, high transmission solar glazing, and all copper absorber plate using black chrome selective coating. *Ametek Power Systems Group. Circle No. 188.*

THE NOVA 4494. A 4'x8' flat plate solar collector for domestichot water. All copper absorber plate: semi-selective surface; low iron tempered glass; NOVA's exclusive EPDM frame. Designed for the economy-minded buyer, but still all high-quality components. EPDM frame won't rust, corrode, crack, chip or peel — ever! Installation kits available. *American Home Solar Energy Systems, Inc. Circle No. 198.*

SOLAR HOME DESIGNS. Home builders looking for readily available passive solar home designs now have them. The Mid-American Solar Energy Complex has made available a booklet of over 15 designs of passive solar homes already built in the Midwest. The homes range in size from 1,200 to 2,000 square feet. The booklet is available now and low cost construction drawings are available for all designs. *Solar 80™ Home Designs. Circle No. 211.*

Specialties

HOW TO WATERPROOF MASONRY WALLS. A 16-page illustrated booklet that explains the common causes of seepage and suggest techniques and products for treating interior and exterior masonry surfaces to withstand moisture penetration. *United Gilsonite Laboratories. Price 25¢. Circle No. 456.*

HOME SUMP PUMP KIT. The "Home Sump Pump Kit" offers information on the selection, installation and maintenance of sump pumps. It includes an installation guide, a trouble shooting chart a homeowner's notebook and a maintenance chart. *Sump Pump Manufacturers Assoc. Price $1.50 Circle No. 483.*

DRY WOOD DOES NOT DECAY. Brochure describes method of prevention and correction of "moisture blistering" of house paint and "dry-rot" fungus control. Easy to install directions. *Midget Louver Co. Price $1.00. Circle No. 452.*

THE FINISHING TOUCH. A 16-page fully illustrated beginner's guide to wood finishing from UGL, manufacturer of ZAR wood finishing products. The booklet takes readers through a typical wood finishing project step-by-step with tips on how to get professional-like results. *United Gilsonite Laboratories. Price 25¢. Circle No. 459.*

UNDERSTANDING UNDERGROUND WATER. This booklet describes the benefits of individual water systems, discusses types of wells and water system components, and tells how to size a system for your home. *Water Systems Council. Price 45¢. Circle No. 455.*

ENERGY-SAVING PRODUCTS. Literature shows how to find energy-wasting gaps around doors and windows and how to seal them with weatherproofing products. Saving energy reduces fuel bills both in winter and in summer. *Myro, Inc. Circle No. 171.*

WATER SYSTEMS PLANNING KIT. Includes "Understanding Underground Water" booklet describing benefits of private wells and water systems, "Order Your Water Well Done", a guide to locating and constructing private water wells and individual manufacturer's product information. *Water Systems Council. Price $1.00. Circle No. 453.*

The Product Information Source

DECORATING WITH CEDAR. A design kit is available from Shakertown to assist the home do-it-yourselfer in the planning and use of Shakertown Fancy Cuts decorative cedar shingles. The new kit includes an idea folder showing creative uses for interior and exterior application, graph paper and tracing guide for making to-scale drawings, a dealer listing, color brochure with product information, and instructions. The tracing guide features 9 varieties of available Fancy Cuts shingle patterns in $1/12$ scale. *Shakertown Corp. Price $1.00. Circle No. 474.*

GO DUCTLESS! Want to help save on construction and energy costs? Go ductless! The Good-Air Ductless Fan is the proven indoor air treatment system that filters contaminants, deodorizes and recirculates cleaner, fresher air continuously. It offers less design restriction, reduces the danger of fire and smoke spreading through ductwork, eliminates exhaust fans, duct-work and outside venting. Conserves energy, too. *Rush-Hampton Industries, Incorp. Circle No. 199.*

PURE DRINKING WATER. Brochure describes residential drinking water purification device as the modern replacement for bottled water. SEAGULL IV removes a wide variety of contaminants including bacteria, TCE, chlorine, and asbestos fibers, making ordinary tap water taste fresh and delicious. "Guide to Choosing a Water Purifier/Filter". *General Ecology Inc. Circle No. 202.*

ORDER YOUR WATER WELL DONE. A guide to locating and constructing water wells. Water quality control is also discussed. *Water Systems Council. Price 75¢. Circle No. 454.*

WATER FILTER. Literature describes remarkable unit that removes sediment and dirt from all household water. Installs easily on the main cold water line. Valve is built into the head of the filter. The unit has three modes of operation: on, off and bypass, permitting quick, easy change of cartridges. Fabricated completely of durable, corrosion free, high impact molded plastic. *Filterite Corp. Circle No. 172.*

SPIRAL STAIRS. Steel and aluminum model spiral stairs for residence or commercial use. Free-standing stair with quality workmanship. Brochure shows steel models pre-finished dark bronze, baked-on enamel. Can be installed in two hours. No special tools required. *Spiral Manufacturing, Inc. Circle No. 173.*

FUN TO MAKE PICTURE FRAMES. Instructions and illustrations for the do-it-yourselfer on how to make picture frames using standard mouldings. Tells what tools to use and how to build and finish picture frames. *Wood Moulding and Millwork Producers. Price 60¢. Circle No. 470.*

MICRO-LAM HEADERS AND BEAMS. A brochure describing Micro-Lam laminated veneer lumber headers and beams is now available from Trus Joist Corp. Stiffer, lighter, longer and stronger than solid sawn lumber, Micro-Lam headers and beams will not shrink, warp, split or twist and have been accepted by all major building code authorities. *Trus Joist Corp. Circle No. 204.*

WATERPROOF COATING. Thoroseal is a cement-base, heavy-duty waterproof coating for all masonry and concrete. Unlike a paint film, Thoroseal permits walls to breathe, prevents damaging accumulation of moisture vapor which forces paint right off the walls. *Thoro System Products. Circle No. 175.*

STORAGE BUILDINGS. Literature shows a complete line of galvanized steel and aluminum storage buildings. The styles include gable, high gable, gambrel and high gambrel in various sizes and colors. They also have a line of all aspenite buildings; all factory cut and drilled. Easy assembly. *Arrow Group Industries. Circle No. 181.*

YORK SPIRAL STAIR. Enhance any home or office with this elegant all wood, double helical stairway. Both inner and outer handrails are provided for safety and comfort. Multi-laminated curved sections with solid treads add strength and long term durability. Available in oak with diameters of 5' and 8'6". Heights are customized. *York Spiral Stair. Circle No. 203.*

RESIDENTIAL TJI ROOF/FLOOR JOIST. A new brochure explaining the efficiency of Residential TJI roof and floor joist. Though lightweight and easy to handle, the "I-shaped" joist is engineered for wider on-center spacings than conventional joists, while offering greater carrying capacities. The TJI joist is available in continuous lengths to 56 feet. *Trus Joist Corp. Circle No. 205.*

SAVE ENERGY & MONEY! 4 unique ways to save money & energy on your water bills. The need for water/energy conserving products is becoming critical. The Water Warden, Moby Dike, Loo Shower Head, Conservarator, and WS-1 Shower flow control provides the solution to this crisis. *Ecology Products Plus, Inc. Circle No. 174.*

Security

MEET ALLSTATE . . . THE GOOD HANDS PEOPLE . . . A handy "Get Acquainted" booklet. Highlights insurance and family financial products you should know about. *Allstate Insurance Co. Circle No. 212.*

SUPER SECURITY. Latch-Gard II, a security device for homes, apartments, motels and hotels, offers real security. A door can be opened safely one inch to view the would-be intruder. The super tough aircraft cable cannot be violated as easily as the common chain door guard. *Latch-Gard Division of Air-Flo Co., Inc. Circle No. 180.*

BEAUTIFUL SECURITY. The Weiser lock booklet illustrates a full range of door knobs, handlesets, levers, and deadbolts to suit every decorating need. Offering a selection of features and functions with finishes that span the spectrum from the shine of bright chrome through the mellow glow of antique etched brass. *Weiser Lock. Price $1.00 Circle No. 482.*

KEYLESS ENTRY. Pushbutton combination door locks. Press a four digit combination from the outside and door unlocks automatically. 10,000 possible combinations. Mechanically activated, no electricity. Interior unlocks by pressing a button. With bilt-in nitelatch, 1" throw deadbolt defies forcible entry. No keys to lose or misplace. Guaranteed pickproof! *Presto-Matic Lock Co. Circle No. 177.*

MAXIMUM SECURITY. Full-color brochure shows complete line of home safes. Complete protection for your valuables. Free standing, wall and floor styles available. *Meilink Safe Co., Div. of Meilink Indust. Inc. Circle No. 178.*

FIRE PROTECTION. Catalog gives information on complete line of sprinklers, nozzles, valves and accessories available in the automatic sprinkler industry. *Grinnell Fire Protection Systems Co., Inc. Circle No. 176.*

HOUSEHOLD INVENTORY RECORD. This folder provides a handy room-by-room inventory record for your personal property. Includes tips on identifying lost or stolen property and a place to record where important papers are kept. *Allstate Insurance Co. Circle No. 212.*

Literature order form

Order the Information You Want Today . . .

STEP I.
Circle the numbers corresponding to the literature in this publication that you want. Please enclose $1.00 processing fee.

FREE BROCHURES & CATALOGS:

101	102	103	104	105	106	107	108
109	110	111	112	113	114	115	116
117	118	119	120	121	122	123	124
125	126	127	128	129	130	131	132
133	134	135	136	137	138	139	140
141	143	145	146	148	149	150	151
152	153	154	156	158	159	160	161
162	163	164	165	167	168	169	170
171	172	173	174	175	176	177	178
180	181	182	183	184	185	186	187
188	189	191	192	194	195	196	197
198	199	202	203	204	205	206	207
208	209	210	211	212	213	214	215
216							

PRICED LITERATURE:

401 25¢	402 50¢	403 25¢	404 75¢	405 $1	406 $1
407 $1	408 50¢	409 75¢	410 25¢	411 60¢	412 25¢
413 25¢	414 $1	415 $1.95	416 $2.95	417 $1	418 $3
419 $1.50	420 75¢	421 75¢	422 75¢	423 $1	424 $1
425 25¢	426 25¢	427 25¢	428 $1	429 $1	430 $20
431 $1	432 $3	433 25¢	434 $2	435 25¢	436 $1
437 $2	438 $1	439 $1	440 30¢	441 25¢	442 $2.50
443 50¢	444 $1	445 $3	446 $1	447 $1	448 10¢
449 10¢	450 25¢	451 35¢	452 $1	453 $1	454 75¢
455 45¢	456 25¢	457 75¢	458 $2	459 25¢	460 $1
461 $1	462 $1	463 $3	464 $1	465 50¢	466 50¢
467 $1	468 $2	469 50¢	470 60¢	471 50¢	472 50¢
473 $4	474 $1	475 50¢	476 $1	478 35¢	479 $1
480 $2	481 50¢	482 $1	483 $1.50		

Complete Order Form Next Page . . .

Literature order form/continued

STEP II.   Please help us out by answering these questions:

1. Are you:
- ☐ Building a new home
- ☐ Buying a new home
- ☐ Buying an older home
- ☐ Remodeling your present home

2. If you are building a new home . . .

A. Are you planning to start construction within the next six months?   ☐ yes ☐ no

B. Do you presently own the land that you plan to build on?   ☐ yes ☐ no

C. How will your home be built?
- ☐ Through a general building contractor
- ☐ With you acting as general contractor
- ☐ With you doing most of the actual labor

D. Where are you getting the construction drawings for your new home?
- ☐ Garlinghouse or another home plans company
- ☐ Architect or local designer
- ☐ Provided by your building contractor
- ☐ Other __________

3.A. Which best describes the size of the closest community or population center to where you live:
- ☐ Under 10,000 population
- ☐ 100,000 to 500,000 population
- ☐ 10,000 to 50,000 population
- ☐ 500,000 to 1,000,000 population
- ☐ 50,000 to 100,000 population
- ☐ over 1,000,000 population

B. Do you live:
- ☐ Within the above community or in its immediate suburb
- ☐ In a rural area, some distance from the above community

4. What is your approximate age?
- ☐ 18-24
- ☐ 50-64
- ☐ 25-34
- ☐ 65 or older
- ☐ 35-49

5. What is the age of your youngest child living at home?
- ☐ No children at home
- ☐ 10-18 years
- ☐ Under 3 years
- ☐ over 18 years
- ☐ 3-9 years

6. What products do you plan to have in your new home?
- ☐ Insulated (double or triple paned) windows and glass door
- ☐ Fireplace and/or wood burning stove
- ☐ Automatic garage door opener
- ☐ Central Air Conditioning
- ☐ Skylights
- ☐ Microwave Oven
- ☐ Dishwasher
- ☐ Washer and Dryer
- ☐ Heat Pump
- ☐ Wall Paneling

7. Are you or a member of your family professionally involved in the building field? If so, please indicate how:
- ☐ No involvement
- ☐ Building contractor
- ☐ Building sub-contractor
- ☐ Architect
- ☐ Home designer (other than architect)
- ☐ Lumber dealer
- ☐ Other building material supplier
- ☐ Engineer involved in residential housing
- ☐ Other __________

______

Step III.   GHPG-2

Fill in your proper mailing address:

Name __________

Address __________

City __________

State __________ Zip __________

Step IV.

Figure the amount due and enclose a check or money order

Amount due for priced literature $ __________

Processing Fee $ ______ 1.00

Kansas Residents Add 3½% Sales Tax $ __________

TOTAL AMOUNT ENCLOSED $ __________

Allow 3-6 weeks for delivery.

______

Step V.   Make checks payable and mail complete page to:

The Garlinghouse Company

P.O. Box 2648, Clinton, Iowa 52735

ONLY PRODUCT LITERATURE MAY BE ORDERED FROM THIS ADDRESS. TO ORDER BLUEPRINTS SEE PAGES 108-109. TO ORDER BUILDING BOOKS SEE PAGE 112.

More Home Plans

Master bedroom suite accentuates luxury

No. 9870—Adorned with pillars and a bow window, this rich French Provincial design becomes an exercise in elegance, crowned by the master bedroom suite. Placed to allow full privacy, the master bedroom incorporates a segmented bath, large walk-in closet, and sitting room with its own closet. A firelit living room and dining room augment an appealing family room, which opens to the terrace. Beyond the kitchen, a laundry room, half bath, and closet space add to the convenience.

First floor—2,015 sq. ft., Basement—2,015 sq. ft.
Garage—545 sq. ft.

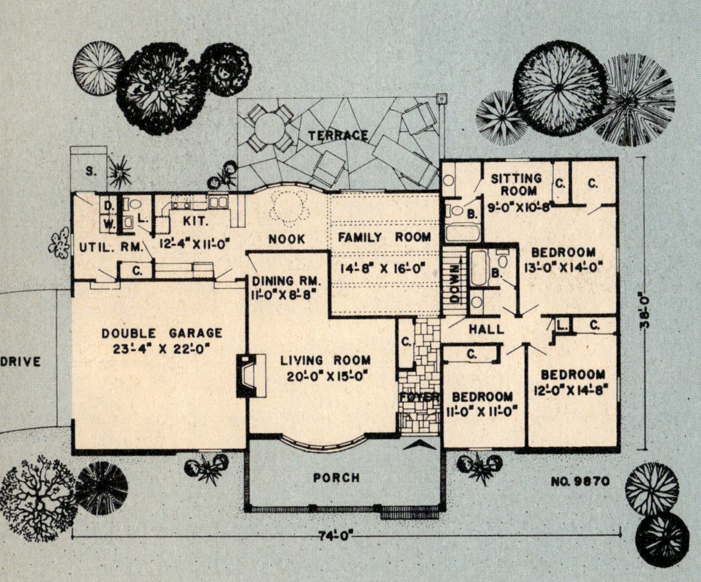

*For price and order information
see pages 108-109.*

Exterior promise of luxury fulfilled

No. 9998—Graceful Spanish arches and stately brick suggest the right attention to detail that is found inside this expansive three bedroom home. The plush master bedroom suite, a prime example, luxuriates in a lounge, walk-in closet and private bath. Exposed rustic beams and a cathedral ceiling heightens the formal living room, and an unusually large family room savors a wood-burning fireplace. In addition to the formal dining room, a kitchen with dinette and access to the terrace is planned.

**First floor—2,333 sq. ft., Basement—2,333 sq. ft.
Garage—599 sq. ft.**

Screened porch expands dining area

No. 9088—Skirting the open kitchen and dining area of this brick-sheathed design, the screened side porch presents an airy setting for family dining. A built-in barbecue serves the kitchen and family room, which also opens to a sprawling flagstone terrace. Bedrooms are sizable and baths efficient, especially the hall bath, an elongated double sink arrangement. Traffic is routed via the comfortable foyer and with coat closet, and the generous living and dining room savors a wood-burning fireplace.

First floor—1,776 sq. ft., Basement—1,776 sq. ft.

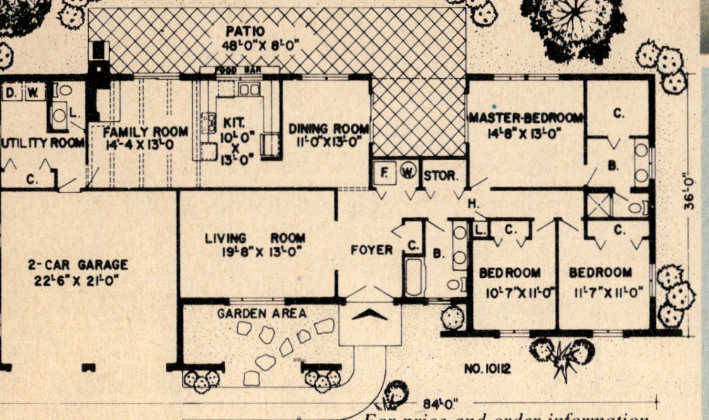

For price and order information
see pages 108-109.

Spanish ranch house offers framed garden

No. 10112—To the left of the foyer, the living room has only one entrance and may easily be reserved for more formal occasions. The large patio, on the other hand, fairly invites use; with sliding door entrances from both the family room and dining room, food bar from the kitchen, and a barbecue, it encourages outdoor living. Three bedrooms are located in a wing of the house isolated from other traffic. The master bedroom makes use of sliding door access to the patio, and a huge walk-in closet.

First floor—2,004 sq. ft., Garage—514 sq. ft.

Living room set apart for formality

No. 9930—Insulated from traffic and activity centers, the expansive living room in this stone-spread ranch style offers a formality and quiet of its own. In the family room, a fireplace engenders a bright and warming atmosphere that can spill out to the terrace via sliding glass doors. A small dining area is situated beyond the kitchen and borders a utility room and half bath. Outstanding in the sleeping wing is the luxury-lined master bedroom with a bath that incorporates "his" and "hers" sinks and a towel closet.

First floor—1,915 sq. ft., Basement—1,915 sq. ft.
Garage—531 sq. ft.

99

Living room highlights ranch style

No. 10176—Dramatically punctuated by a wood-burning fireplace on one end and glass access to the terrace on the other, the expansive living room becomes the focus of this distinctive ranch style. Two closets furnish the long foyer, and, to the right, three bedrooms and two full baths are nestled in the sleeping wing. Behind the double garage, the utility complex merits a laundry center and full bath with shower, while the large kitchen offers informal dining space.

**First floor—1,608 sq. ft., Basement—1,608 sq. ft.
Garage—576 sq. ft.**

Porch connects garage, foyer

No. 10212—Accessibility is a prime consideration in this trim three bedroom traditional, where the covered porch connects to the entry to provide shelter, a link to the garage, and a warm welcome. Within steps of the foyer is a formal bow-windowed living room and a well-placed corridor kitchen. Laundry and half bath border the kitchen on one end, while an airy family room at the other end boasts a dining area, storage space, and sliding glass doors to the patio.

**First floor—2,055 sq. ft., Basement—2,055 sq. ft.
Garage—552 sq. ft.**

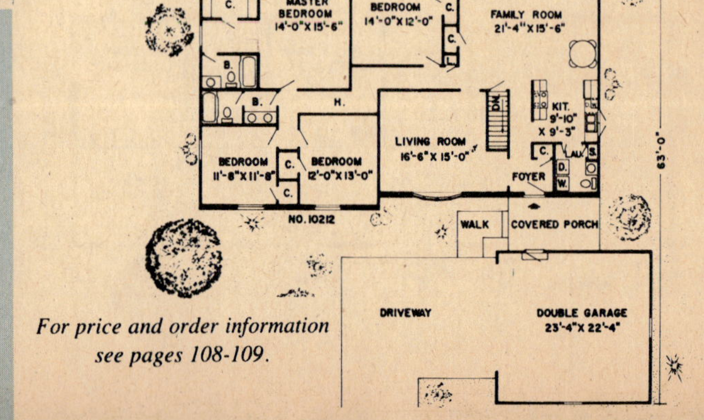

*For price and order information
see pages 108-109.*

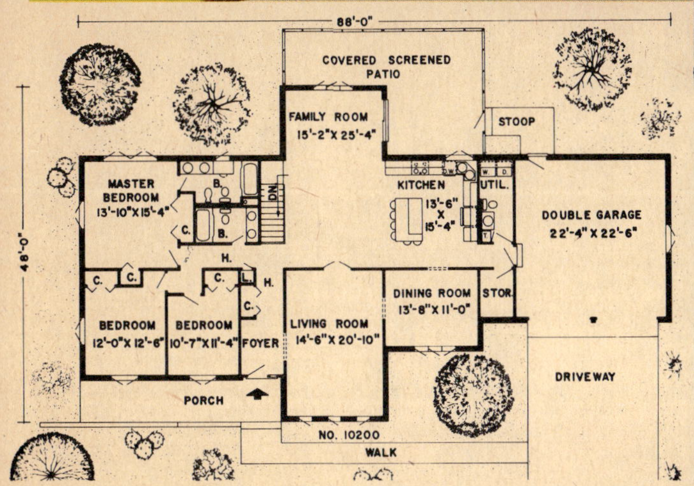

Airy ranch style features screen patio

No. 10200—Open to the family room and steps from the kitchen for cookouts, the covered screened porch provides the perfect accent for this spacious and well-windowed contemporary ranch style. Central foyer and hallway channel traffic to bedrooms, formal living and dining rooms, and the expansive family room-kitchen complex. Serving as a focal point of the plan, the family-kitchen borders a utility room and shows entrances to back yard and garage. Three bedrooms and two full baths make up the sleeping wing.

**First floor—2,254 sq. ft., Basement—2,134 sq. ft.
Garage—554 sq. ft.**

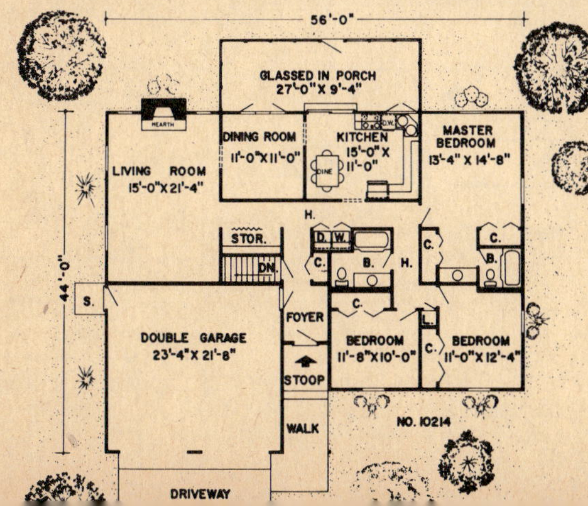

Design focuses on formality

No. 10214—Situated to overlook an impressive 27-ft. glassed-in porch, the dining room joins the formal living room of this ranch plan to create a workable unit for formal entertaining. A cozy fireplace furnishes the living room, and the large kitchen reserves space for family dining. Sizable and well-windowed, the master bedroom offers double closets, dressing area and full bath, and another full bath serves two more bedrooms. For convenience, the double garage opens directly into the foyer.

**First floor—1,651 sq. ft., Basement—1,651 sq. ft.
Garage—521 sq. ft.**

Three fireplaces decorate home

No. 9514—A fireplace unit provides corner fireplaces for the living and family rooms of this ranch home. The living room incorporates a dining area opening onto the kitchen, complete with breakfast nook. Opening from the garage is a shop or hobby area. Three bedrooms with one and one-half baths complete the sleeping area. In addition to large closets in the bedrooms, a closet in the mudroom and hall and a linen closet in the bathroom give adequate storage room.

**First floor—1,726 sq. ft., Basement—1,726 sq. ft.
Garage and shop—558 sq. ft.**

An exciting plan that's full of great features

No. 196—Angled inward to provide relaxation for both master bedroom and family room, the covered patio joins with the screened porch in creating sheltered outdoor offshoots of this contemporary plan. Privacy characterizes the patio, open to the elegant master bedroom with bath and dressing room, while the screened porch adjoins the kitchen and suggests enjoyable outdoor dining. Extending from the kitchen is the airy family room furnished with breakfast bar and fireplace.

**First floor—1,805 sq. ft., Basement—970 sq. ft.
Garage—475 sq. ft.**

This is beautiful

No. 9540—This Ranch-type house features maximum livability. Four large bedrooms and three full baths are included. The entry is very attractive with the built-in planter serving as a divider between the living room and hall. The cut stone fireplace divides the living room and dining room. The hall bath is compartmented and includes a separate shower and built-in dressing table. The kitchen is large and roomy and contains a built-in range and oven and space for the washer and dryer.

First floor—2,616 sq. ft., Basement—1,897 sq. ft.
Garage—528 sq. ft.

Great living area

No. 9542—Picturesque and warmly inviting, this California style Ranch house will be very easy to live in. The floor plan arrangement is well designed. The center hall and foyer provide excellent circulation. The living room has many furniture arrangement possibilities. The master bedroom has its own bathroom with an attractive square tub. All bedrooms contain generous sliding door wardrobe closets. A full basement is provided for utilities, storage, recreation room, laundry, etc.

First floor—1,708 sq. ft., Basement—1,708 sq. ft.
Garage—590 sq. ft.

*For price and order information
see pages 108-109.*

Private deck for sunning or relaxing

No. 9980—Mediterranean styling is quite prevalent in this beautiful family home. The exterior features a walled veranda or courtyard entrance, arched windows, exposed rafter ends and projecting beams made of rough cedar, shake shingled roof and antique brick. At the rear is a wood deck which wraps around the corner, thereby providing a shady area most of the time. Sliding glass doors in the family room and den open onto the deck. The den is intended to be a sitting room for the master bedroom as well as a den.

First floor—2,264 sq. ft., Basement—2,264 sq. ft.
Garage—615 sq. ft.

Simplified entertainment center

No. 9932—This beautiful home possesses individuality and character. This is accomplished by combining the mansard roofs with a steep hip roof, and using clinker brick and battened rough cedar siding for exterior finishes. These materials combined with the shake shingled roof produce a striking exterior. The floor plan is excellent. The master bedroom suite has a dressing area, walk-in closet and a compartmented private bath.

First floor—2,315 sq. ft., Basement—2,315 sq. ft.
Garage—537 sq. ft.

Colonial detailing enlivens exterior

No. 10020—Impressive Colonial columns punctuate the semi-circular porch and fuse with the bow windows and brick to create an exceptional facade. Inside, the floor plan is a study in modern living. Fireplaces grace both living room and family room, which opens to an expansive terrace. A formal dining room adjoins the highly functional kitchen, and the 21 foot master bedroom boasts a lavish full bath and double closets. Two front bedrooms are accented with lovely bow windows.

**First floor—2,512 sq. ft., Basement—2,512 sq. ft.
Garage—648 sq. ft.**

Abundance of space enhances livability

No. 9886—Generously proportioned rooms and open planning result in a considerable amount of space for living in this sleek, brick trimmed design. To the right of the tiled entry, a combination living and dining room measures over 26 feet long, and annexes as even larger family room and kitchen complex with access to the terrace. Even the laundry room is sizable and houses a sink and storage closet. The well planned bedroom wing allots three bedrooms, two full baths, and plentiful closet space.

**First floor—1,881 sq. ft., Basement—1,678 sq. ft.
Garage—590 sq. ft.**

Cupolas and ornamental iron add prestige

No. 9944—A home like this one retains its resale value indefinitely. The stone veneer construction provides a low maintenance exterior and is extremely attractive. The floor plan features three large bedrooms plus a den, two full baths, a powder room near the front entrance and a half bath near the den and laundry. The centrally located kitchen opens into the family room. A breakfast bar divides the two rooms. The family room has a wood burning fireplace.

First floor—2,422 sq. ft., Basement—2,422 sq. ft.
Garage—533 sq. ft.

Enclosed garden provides privacy

No. 10052—A private world for parents is the dominating feature of this beautiful home. This area has a sitting room with fireplace, a large bedroom, two large walk-in closets and a compartmented four-piece bath. The rest of the house contains a family room with fireplace, a large living room, a well equipped kitchen, formal dining room, three bedrooms, a large bath and a laundry room.

First floor—2,577 sq. ft., Basement—1,928 sq. ft.
Garage—798 sq. ft., Private garden—105 sq. ft.

Living room sparkles with light, warmth

No. 9946—Cathedral ceilings terminating in gable end windows blend with rough cedar beams and a wood-burning fireplace to design a unique living room, alive with light and atmosphere. Natural stone, shake shingles and rough cedar layer the exterior of this singular contemporary. Inside, the living room is nestled near bedrooms for quiet, while activity centers fill the area behind the foyer. Open family room and kitchen enjoy breakfast bar and wooden deck, while a compact laundry and half bath separates kitchen and dining room.

First floor—2,096 sq. ft., Basement—2,096 sq. ft.
Garage—624 sq. ft.

*For price and order information
see pages 108-109.*

Angle set for entertaining, relaxing

No. 10124—Striking and expensive, this three bedroom ranch style welcomes its angled plan as a means of effectively zoning living areas. To the left of the foyer, bedrooms are nestled around two full baths and are well-closeted for comfort. The bordering living room helps buffer noise and offers a quiet spot for relaxing. Entertaining is encouraged in the large, angular family room, which sports fireplace, bar, and sliding glass doors to the patio.

First floor—2,180 sq. ft.
Garage and storage—580 sq. ft.
Basement—2,042 sq. ft.

ORDER YOUR NEW

From a compan

- Experience the thrill of creating and customizing your home.
- Enjoy comparing and choosing options.
- Obtain construction bids . . . and you're ready to build.

For 75 years, Garlinghouse homes have been built by tens of thousands of families across the nation. The construction blueprints are complete ... accurate ... and contain all the information a builder needs to begin construction.

GARLINGHOUSE PLANS SAVE YOU MONEY

The costs of designing our homes are spread over a number of plan buyers, nationwide. Therefore, you pay only a fraction of what you would spend to have a home designed (specifically for you).

BLUEPRINT MODIFICATIONS

It is expensive to have plans modified by professional designers. However, minor alterations in design, as well as building material substitutions, can be made by any competent builder according to the needs or wishes of the home owners.

YOU CAN CHARGE YOUR ORDER!

TO ORDER PRODUCT LITERATURE SEE PAGES 95-9

BLUEPRINT ORDER FORM

05045

PLEASE SEND ME:
- ☐ One Complete Set of Blueprints
- ☐ Minimum Construction Package five sets
- ☐ Standard Construction Package eight sets

01100

Plan Number _____ ☐ as shown ☐ reversed

Cost . . . . $_____

    Additional Set(s) $15.00 each . . . . $_____

Materials List ($10.00 per order) . . . . $_____

Mailing Charges . . . . $_____ 3.50

TOTAL AMOUNT ENCLOSED . . . . $_____
(Kansas residents add 3 1/2%)

CHARGE MY ORDER TO:
- ☐ Mastercharge ☐ Visa ☐ American Express
  Exp.

Card # _____ Date _____

Signature _____

WE WOULD APPRECIATE YOUR HELP IN ANSWERING THE FOLLOWING QUESTIONS:

	Yes	No
Do you now own the land you plan to build on?	☐	☐
Do you plan to start construction in the next 6 months?	☐	☐
Do you plan to build most of the home yourself?	☐	☐
Would you like for us to have free building product information sent to you?	☐	☐

Name _____

Address _____

City _____ State _____ Zip _____

The Garlinghouse Co., P.O. Box 299, 320 S.W. 33rd St.
(913) 267-2490   Topeka, Kansas 66601

BLUEPRINT ORDER FORM

05045

PLEASE SEND ME:
- ☐ One Complete Set of Blueprints
- ☐ Minimum Construction Package five sets
- ☐ Standard Construction Package eight sets

01100

Plan Number _____ ☐ as shown ☐ reversed

Cost . . . . $_____

    Additional Set(s) $15.00 each . . . . $_____

Materials List ($10.00 per order) . . . . $_____

Mailing Charges . . . . $_____ 3.50

TOTAL AMOUNT ENCLOSED . . . . $_____
(Kansas residents add 3 1/2%)

CHARGE MY ORDER TO:
- ☐ Mastercharge ☐ Visa ☐ American Express
  Exp.

Card # _____ Date _____

Signature _____

WE WOULD APPRECIATE YOUR HELP IN ANSWERING THE FOLLOWING QUESTIONS:

	Yes	No
Do you now own the land you plan to build on?	☐	☐
Do you plan to start construction in the next 6 months?	☐	☐
Do you plan to build most of the home yourself?	☐	☐
Would you like for us to have free building product information sent to you?	☐	☐

Name _____

Address _____

City _____ State _____ Zip _____

The Garlinghouse Co., P.O. Box 299, 320 S.W. 33rd St.
(913) 267-2490   Topeka, Kansas 66601

HOME BLUEPRINTS ... TODAY!!
ith 75 years of experience.

WHAT IS INCLUDED IN A SET OF GARLINGHOUSE BLUEPRINTS?

1) Four Elevations (Views of the Outside)
2) Floor Plans (1/4" to 1' scale) for all floors
3) Basement and/or Foundation Plan
4) Roof Plan
5) Typical Wall Sections
6) Kitchen Cabinet Details
7) Fireplace Detail (where applicable)
8) Detail for Stairs (where applicable)
9) Plot Plan
10) Electrical Layout
11) Complete Materials List (If Ordered)
12) Specifications & Contract Form

REVERSE PLANS

You may find that a particular house would suit you or your lot condition better if it were reversed. A reverse plan turns the design end for end. That is, if the garage is shown on the left side and the bedrooms on the right, the reverse plan will show the garage on the right side and the bedrooms on the left. To see how a design will look in reverse, hold the book in front of a mirror. Dimensions and lettering of most Garlinghouse reverse plans are right reading. When this is not the case, one mirror image set is produced for reference by you and your builder, and the rest are sent "as shown" for ease of reading lettering and dimensions. (Available only on multiple set orders.)

MONEY BACK GUARANTEE

All Garlinghouse Blueprints are sold with a ten day unconditional money back guarantee. Study the plan sets for ten days ... even get an opinion from your builder. If you are not completely satisfied, then return the blueprints and your money will be cheerfully refunded ... no questions asked.

INTERNATIONAL ORDERS

If you are ordering from outside the United States, then your check, money order, or international money transfer should be payable in U.S. currency. Because of the extremely long delays involved with surface mail, we ship all orders from outside the U.S.A., Canada or Mexico via Air Mail. There is a total mailing charge of $18.00 on these orders.

Canadian and Mexican orders are normally shipped by first class mail. Please include $5.00 for mailing charges on these orders. If you wish to have your order shipped into Mexico via Air Mail, then you will need to enclose $9.00 to cover the mailing charges.

All blueprints orders are shipped UPS or Priority Mail (usually on the same day the order is received).

An additional flat charge of $10.00 is required to have a materials list included with each set of plans ordered.

Prices are subject to change without notice.

BLUEPRINT ORDER FORM

01100

05045

PLEASE SEND ME:

- ☐ One Complete Set of Blueprints
- ☐ Minimum Construction Package five sets
- ☐ Standard Construction Package eight sets

Plan Number _______ ☐ as shown ☐ reversed

Cost .... $ _______

    Additional Set(s) $15.00 each .... $ _______

Materials List ($10.00 per order) .... $ _______

Mailing Charges .... $ 3.50

TOTAL AMOUNT ENCLOSED .... $ _______
(Kansas residents add 3 1/2%)

CHARGE MY ORDER TO:

☐ Masterchange ☐ Visa ☐ American Express

Card # ___________________________   Exp. Date _______

Signature ___________________________

WE WOULD APPRECIATE YOUR HELP IN ANSWERING THE FOLLOWING QUESTIONS:

	Yes	No
Do you now own the land you plan to build on?	☐	☐
Do you plan to start construction in the next 6 months?	☐	☐
Do you plan to build most of the home yourself?	☐	☐
Would you like for us to have free building product information sent to you?	☐	☐

Name _______________________________

Address _____________________________

City ____________ State ______ Zip ______

The Garlinghouse Co., P.O. Box 299, 320 S.W. 33rd St.
(913) 267-2490   Topeka, Kansas 66601

NOTICE

One or more of the books you have ordered may not contain current plan prices. Please use the schedule below when ordering plans:

SINGLE FAMILY HOMES	1 set	5 sets	8 sets	dup. sets
Under 1400 sq. ft.	$60.00	$95.00	$120.00	$15.00
1400 to 2400 sq. ft.	65.00	100.00	125.00	15.00
Over 2400 sq. ft.	70.00	105.00	130.00	15.00
DUPLEX DESIGNS	80.00	115.00	140.00	15.00
TRIPLEX or LARGER	95.00	130.00	155.00	15.00

(Please add $4.25 for postage and handling on all orders.)

An additional flat charge of $10.00 is required to have a materials list included with each set of plans ordered.

We require **full** payment to accompany all orders. To avoid the delay of having your order returned for additional money, please refer to the above schedule when preparing your check or money order. For those customers wishing to use a credit card to charge their order, we accept VISA, MasterCard, and American Express.

The Garlinghouse Co. • P.O. Box 299 • Topeka, KS 66601 • 913-267-2490

Let Us Help You Save Dollars In Fuel Costs!

01100

We don't need to tell you about high energy costs — you already know that . . .

But did you know that many peoples fuel bills equal or exceed their house payment? Or that fuel prices are expected to TRIPLE within the next ten years?

At Garlinghouse, we care, and we want to help you save those fuel dollars in your new home! In fact, we feel this is so important, that we have prepared a new **Energy Conservation Specification Guide** for our customers. And it's FREE with every blueprint order!

Here in one complete easy-to-understand guide learn:

- About 22 fuel-saving measures **you** can use!
- How home location affects fuel costs.
- About the different types of insulation available and what they can do for you.
- How a few details can go a long way in saving energy!
- How, with the aid of our insulation spec. charts to plan your insulation needs.

It's all yours — FREE

"""

Building Books

ORDER FORM
PAGE 112

The books on this page were written with the professional home builder in mind. They are all comprehensive information sources for contractors or for those beginners who wish to build like contractors.

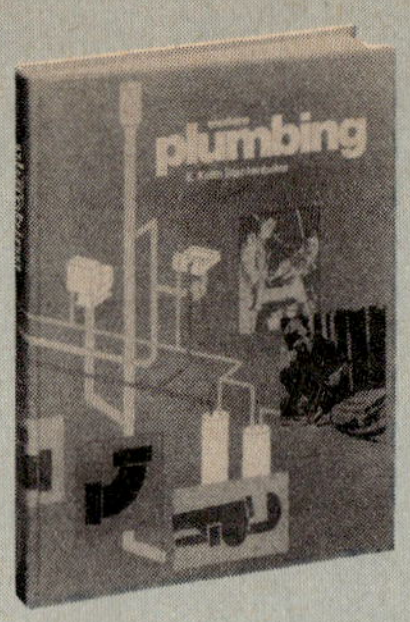

2508. Modern Plumbing All aspects of plumbing installation, service, and repair are presented here in illustrated, easy-to-follow text. This book contains all the information needed for vocational competence, including the most up-to-date tools, materials, and practices. 300 pp.; over 700 illus. Goodheart-Willcox **$14.00**

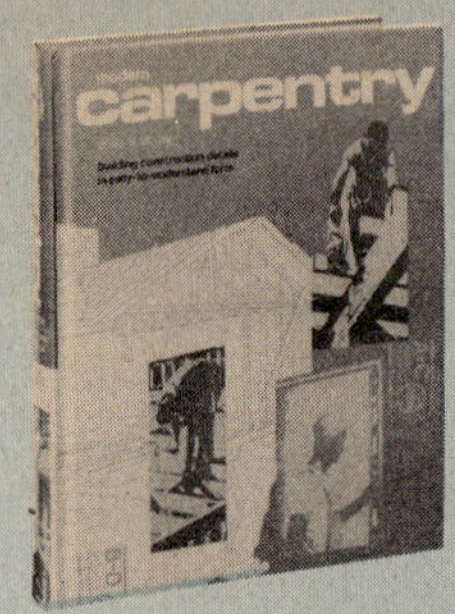

2510. Modern Carpentry A complete guide to the "nuts and bolts" of building a home. This book explains all about building materials, framing, trim work, insulation, foundations, and much more. A valuable text and reference guide. 492 pp.; over 1400 illus. Goodheart-Willcox **$14.00**

2506. House Wiring Simplified This book teaches all the fundamentals of modern house wiring; shows how it's done with easy-to-understand drawings. A thorough guide to the materials and practices for safe, efficient installation of home electrical systems. 176 pp.; 384 illus. Goodheart-Willcox **$8.00**

2512. Complete Concrete, Masonry, and Brick Handbook This is exactly what it says it is; the largest and most complete book available. Everything that you will ever need to know about working with concrete, masonry, and brick is here, in step-by-step detail. 1072 pp.; over 900 illus. Arco **$19.95**

2500. Goodheart - Willcox's Painting and Decorating Encyclopedia A complete library of professional know-how on painting, decorating, and wood finishing in one easy-to-use volume. Includes "how-to" sections on paint selections; properties and function of the various paint components; proper use of tools and equipment; surface preparation; paint and wall covering application; wood finishing; etc. 272 pp.; 300 illus. Goodheart-Willcox **$9.28**

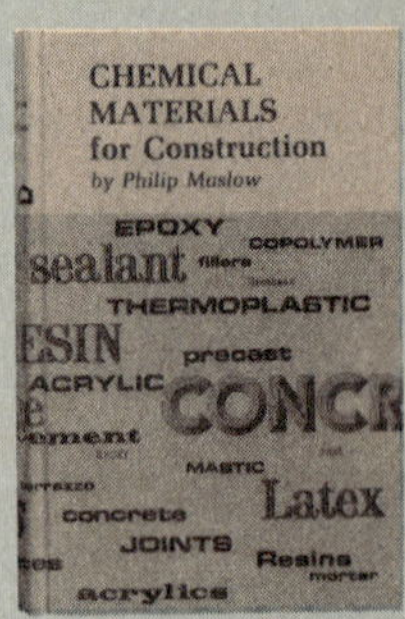

2502. Chemical Materials For Construction The most complete reference available on the properties, applications, and kinds of chemical substances used in construction. This handbook covers concrete and chemical materials used with concrete; caulks and sealants; adhesives; waterproofing materials; roofing; flooring; and more. For the true professional who wants to be sure he has the right chemical substance for the specific job application, this book is a must. 592 pp.; Structures **$32.50**

2514. The Underground House Book For anyone seriously interested in building and living in an underground home, this book tells it all. Aesthetic considerations, building codes, site planning, financing, insurance, planning and decorating considerations, maintenance costs, soil, excavation, landscaping, water considerations, humidity control, and specific case histories are among the many facets of underground living dealt with in this publication. 208 pp.; 140 illus. Garden Way **$9.95**

2544. Solar Houses An examination of solar homes from the standpoint of lifestyle. This publication shows you through photographs, interviews, and practical information, what a solar lifestyle involves, how owners react to it, and what the bottom-line economics are. Included are 130 floor plans and diagrams which give you a clear idea of how various "active" and "passive" solar systems work. 160 pp.; 370 illus. Pantheon (paperback) **$9.95**

2520. Successful Swimming Pools Everything the would-be owner should know — from choosing a pool you can afford to securing permits, purchasing accessories and building related structures. Special sections for the build-it-yourselfer cover construction problems, safety and lighting. Includes a manufacturers list. 152 pp.; 250 illus. Structures (paperback) **$6.95**

2524. Successful Bathrooms Expert advice on all aspects of creating the smallest room in your home. Careful attention is given to lighting, fixtures, fittings, hot tubs, saunas, water-saving devices, safety, and more. Abundant color photographs, unique bathroom floor plans, and a manufacturer's index all help to get your bathroom project off the ground. 136 pp.; 200 illus. Structures (paperback) **$6.95**

2516. Building Consultant The new home buyer's bible to home construction. This encyclopedia of home building explains in comprehensive detail about all the various elements that go into a completed house. It enables you to deal with the construction of your new home in a meaningful way that will avoid costly errors, whether you use a contractor or build it yourself. 188 pp.; Holland House (paperback) **$6.00**

2536. Build Your Own Solar Water Heater A nuts and bolts guide written in clear "how-to-do-it" terms. This book describes only workable, efficient systems that can be built with off-the-shelf plumbing components. Also included is a complete discussion of collectors, financial analysis of solar hot water systems, and even special considerations for solar heated swimming pool systems. 120 pp.; 75 illus. Garden Way (paperback) **$7.95**

2526. Successful Kitchens An excellent planning guide touching on kitchen layout, lighting, cabinets, counter tops, sinks, appliances, and noise control. This book offers solid information on the things you need to know to create the kitchen that best fits your needs. The color photos of various model kitchens are a great help in generating ideas. 144 pp.; Structures (paperback) **$6.95**

2518. Build Your Own Home An authoritative guide on how to be your own general contractor. This book goes through the step-by-step process of building a house with special emphasis on the business aspects such as financing, scheduling, permits, insurance, and more. Furthermore, it gives you an understanding of what to expect out of your various sub-contractors so that you can properly orchestrate their work. 106 pp.; Holland House (paperback) **$6.00**

2528. Successful Fireplaces This publication, in its 20th printing, has sold over 1,000,000 copies. A wealth of information on fireplace construction techniques, safety, energy saving considerations, wood stoves, and the use of wood as a fuel. Contains a variety of interior ideas, including many color photos, to help you determine the right fireplace (or wood stove) for your home. Manufacturers of fireplaces and related equipment are also listed. 129 pp.; Structures (paperback) **$6.95**

2522. How To Build Your Own Home For the person who wants to build a home with his own two hands. This publication explains in generously illustrated detail, each step of the construction process, leaving nothing out. Surveying, excavating, carpentry, plumbing, heating, and resilient flooring are just a few of the many items covered. 356 pp.; Structures (paperback) **$9.95**

Building Books Order Form

PLEASE SEND ME THE FOLLOWING BOOKS:

Book Order Number	Price
_______________	$ _____________
_______________	$ _____________
_______________	$ _____________
_______________	$ _____________
_______________	$ _____________
Postage and Handling	$ _____1.00_____
Kansas Residents Add 3½% Sales Tax	$ _____________
TOTAL ENCLOSED	**$** _____________

- Orders usually shipped the same day they are received
- International Orders include an additional $1.50 per book for surface mail

CHARGE MY ORDER TO:

☐ **MasterCard**   ☐ **Visa**   ☐ **American Express**

Card # _________________________________

Exp. Date _______________

Signature _________________________________

____

MY SHIPPING ADDRESS IS:
(Please Print)

Name _________________________________

Address _______________________________

City ___________________________________

State _______________________ Zip _____________

05046

SEND ORDER TO:

The Garlinghouse Company
P.O. Box 299 - 320 SW 33rd St. 01100
Topeka, Kansas 66601

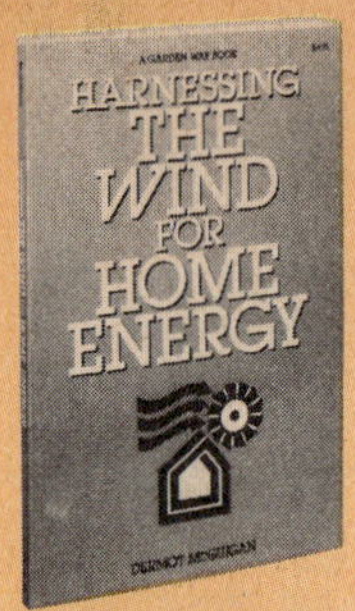

2532. Harnessing The Wind For Home Energy
144 pp. Garden Way (paperback) **$4.95**

2534. Harnessing Water Power For Home Energy
112 pp.; Garden Way (paperback) **$4.95**

Wind and water power! Are they for you? Each of these books includes an in-depth analysis assessing the potential of your home site for using wind or water energy; the various types of systems available and which is best for you; return or payback on investment; and a listing of manufacturers of wind or water power systems. If you have ever considered using wind or water energy, then these are the books for you.

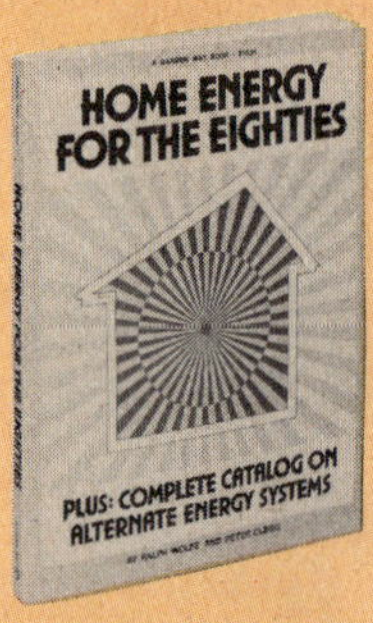

2538. Home Energy For The Eighties Covering solar energy, wind power, water power, wood heat, as well as energy conservation, this book provides a complete discussion of the theory, technology, and equipment in each field. Non-technical explanations and name-brand product descriptions with comparison charts are especially useful. This is perhaps the most practical and thorough book about home energy on the market today. 272 pp.; 300 illus. Garden Way (paperback) **$10.95**

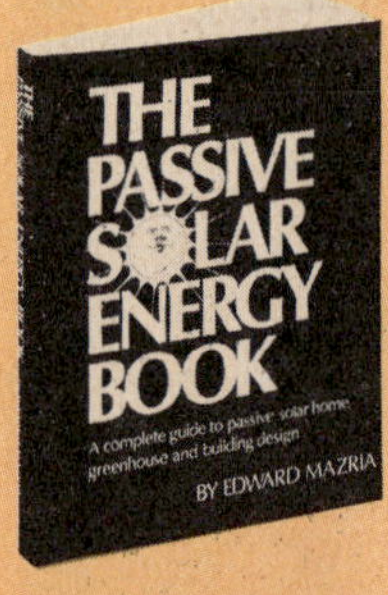

2540. The Passive Solar Energy Book A surprisingly complete guide to passive solar home, greenhouse, and building design. This book presents a step-by-step process for choosing and sizing the systems best suited for your particular needs. Includes information about solar radiation, regional climate variations, and space heat losses and gains so that you can calculate heating/cooling requirements and determine the potential money savings with a passive solar system. 448 pp.; 238 illus. Rodale (paperback) **$12.95**

2542. Designing and Building A Solar House Written by one of America's foremost authorities on solar architecture. It is a practical "how-to" guide that clearly demonstrates the most sensible ways to marry good house design with contemporary solar technology. Included is a thorough discussion of both "active" and "passive" solar systems, and even a listing of the today's leading solar homes. 288 pp.; 400 illus. Garden Way (paperback) **$10.95**

2530. How To Build and Buy Cabinets For The Modern Kitchen A superior publication for the do-it-yourselfer. This book explains all the design considerations that must be kept in mind, and then shows you how to build almost every conceivable type of kitchen cabinet available. Accurate and detailed working drawings with exact measurement insure that even the novice cabinet builder can create a professional job. 269 pp.; Arco **$10.00**